Shadowless Squids

Shadowless Squids
Stories of Physics in Nature

Vitalii Zablotskii
Tatyana Polyakova

JENNY STANFORD
PUBLISHING

Published by

Jenny Stanford Publishing Pte. Ltd.
101 Thomson Road
#06-01, United Square
Singapore 307591

Email: editorial@jennystanford.com
Web: www.jennystanford.com

British Library Cataloguing-in-Publication Data
A catalogue record for this book is available from the British Library.

Illustrator: Olha Vinnik

ISBN 978-981-5129-43-4 (Hardcover)
ISBN 978-1-003-57062-2 (eBook)

Contents

Preface

Do animals think? Do they know the laws of nature and use them in their lives? Who can move without casting a shadow? Do animals possess nanotechnologies and some secret knowledge?

You will find answers to these and other questions by reading this book. You will also learn about the unusual abilities of animals to predict earthquakes and tsunamis, find out why crocodiles have flat eyes and how your cat always manages to land on its feet, how a spider can fly hundreds of kilometers, and in what extreme conditions a cheetah must hunt. You will attend an unusual concert, embark on a fascinating journey inside a living cell, and even come close to a black hole, peering beyond the event horizon, among many other things. In the pages of this book, you will meet a scholarly dog, a cat-philosopher who asks intriguing questions, and a curious and insightful reader (we hope you will be such a reader). Together with the authors, they will explain to you the unusual abilities of animals and some questions of biophysics. You can read this book with your children, explaining simple laws from the school physics course.

If you love animals and are interested in physics and biology, then this book is for you. Here you will find engaging stories that convey interesting knowledge and facts about the behavior, hunting, and lifestyle of wild animals. These stories are written in various literary styles: fairy tales, monologues, plays, and even detective stories. In our stories, we, the authors, often lead the reader to the cutting edge of science and touch upon questions to which science does not yet know the answer, hoping that our young readers will choose a scientific career in the field of physics or biology in the near future and find answers in their scientific research. This book is written precisely for this purpose.

Even if your life is not related to science, this book will still be useful to you. Why? For example, if someone jumps on your head at five o'clock in the morning or looks at you with tea-colored eyes early in the morning, demanding a walk even when it's cold and raining outside, it means you have a four-legged friend. And friends should not only be loved but also understood. We sincerely hope that the knowledge you gain from this book will help you better understand your pets and find common interests with them.

Vitalii Zablotskii
Tatyana Polyakova
Summer 2024

Chapter 1

A Lakeside Concert

"I have two tickets to the frog concert. Shall we go?" he asked.

"Yes, it's so romantic! And where is it taking place?" she replied.

"On a lake near the city. Our seats are right by the water. So, we shall see and hear everything perfectly. Moreover, it is said that a listener of a "'frog concert' is guaranteed career advancement and material well-being."

"I've never been to a frog concert before. What dress should I wear? I think a green evening dress with a generous neckline would be appropriate."

"Of course, green or yellow will match the costumes of our singers. But I think it would be better to cover up and take waterproof shoes. There will definitely be mosquitoes on the lake in addition to the frog performers."

The concert started before sunset. Frogs' voices floated over the water, ranging from soprano to bass.

"Hey, look, there are big white sacs ballooning on the heads of the frogs—the soloists. What is it and what is it for?" she asked.

"Those are male frogs. And they sing to attract the female. Frog singing usually means something like, "I'm here, my love!" Who sings louder and more melodious, the females will be more attracted to them. And the white, periodically inflating throat sacs on the male's head are resonators that serve to amplify the sound."

Shadowless Squids: Stories of Physics in Nature
Vitalii Zablotskii and Tatyana Polyakova
Copyright © 2025 Jenny Stanford Publishing Pte. Ltd.
ISBN 978-981-5129-43-4 (Hardcover), 978-1-003-57062-2 (eBook)
www.jennystanford.com

"I understood everything about love, but I didn't understand anything about the air sacs. What are resonators? And how do they amplify the sound? Can you explain it to me in more detail?"

"Yes. But it's not going to happen without a little physics lesson." (The reader who has a good knowledge of the physics of wave processes can skip this part.)

"Acoustic resonance and resonators. Sound is a wave (periodic compression and expansion) propagating at a certain speed in elastic environments such as air or water. Like all waves, sound waves have the ability to bounce off the interface between two mediums and refract at their boundary. It is important to know that sound waves can add to and amplify or attenuate each other. But when two waves come together, they amplify or cancel each other out only if they are coherent. Waves are then called coherent if they have the same frequency and constant phase difference. Oh, that phase... Few students understand what it is. Mathematically, it's simply the argument of a sine or cosine function that describes harmonics and waves. Waves arriving at a given point in space with the same phase φ (or with a phase difference that is a multiple of the integer 2π) amplify each other so that the amplitude of the resulting wave is doubled. Simply put, waves arriving at a point in space in the same phase arrive such that the maximum of one wave superimposes on the maximum of the other. And waves arriving with opposite phases (or with a phase difference that is a multiple of an odd integer π, or in other words waves arriving at a point in space such that the maximum of one wave is superimposed on the minimum of the other wave) will cancel each other out, i.e., the amplitude of the resulting wave will be zero. The effect of increasing the amplitude when two sound waves are added is used to amplify the sound using resonators."

"Ah... amplification of sound then??? What exactly are these resonators???"

"Resonators are devices designed to amplify acoustic oscillations and increase the amplitude of sound waves emitted by vibrating bodies such as strings. In the simplest case, it is a wooden rectangular box open at one (front) end. The length of the box is based on the condition that the phases of the direct wave and the wave reflected from the rear wall of the resonator coincide: $\Delta\varphi = 2\pi n$, where n is an integer. To satisfy this condition, the length of the resonator must be equal to $L = \lambda n/2$, where λ is the length of the sound wave."

"And... what about an application example, would you have one?"

He took a breath and continued, "For example, a musical tuner on a rectangular box. Here, the length of the box exactly meets the above condition.

The body of a violin or guitar is also an acoustic resonator. The shape of the sound box is made in such a way that all sound wavelengths (λ) emitted by the vibrating violin strings will find the desired distance between the walls of the body: $L = \lambda n/2$. The waves that meet these resonance conditions are amplified, making the violin sounds louder. Do frogs know these resonance conditions? They probably do, because they inflate air sacs near their heads, which increases the amplitude of their croaks. But to us, it just seems like they're croaking. They're actually singing a song. Moreover, the sound repertoire of *Hylodes japi* frogs turns out to be quite diverse: they croak, scream, squeal, etc. In this species of frog, scientists have recorded 18 types of sounds...."

"Fantastic!!!" she gasped.

"But now let's talk about resonance in the language of oscillating frequencies," he said. "Resonance is a sharp increase in the amplitude of oscillations that occurs when the frequency of the driving force coincides with the natural frequency of oscillations of the system. In one account about the famous Italian singer Enrico Caruso, the story goes that once at a banquet Caruso showed off not only his beautiful voice but also his knowledge of physics. He picked up an elegant wine glass from the table, tapped his fingers lightly on the side of the glass and listened to the note the glass wall vibrated to... do you know why?"

"To tune his voice?"

"Actually yes!!!" he nodded cheerfully. "More precisely, to detect the natural frequency of vibrations of the glass wall. He then raised the glass to his mouth and with his voice made a powerful sound of exactly the same frequency (in this case the sound waves of the singer's voice were the driving force for the oscillating walls of the glass). And a miracle happened (which, of course, only those unfamiliar with physics would say): resonance set in, the amplitude of the vibrations of the walls increased sharply, and the glass shattered into small fragments."

"But back to our frogs: female frogs never croak, probably because they are quiet and modest. They might only sputter or make a strange noise if they suddenly feel pain. The male bullfrog has the strongest voice. When people first heard it, they refused to believe that the sound was made by a frog and not a large animal. The male *Metaphrynella* frog is not limited to using the air sac resonator that nature has provided. It looks for a future dwelling—a cavity in the trunk of a tree, evaluates it not only in terms of the convenience of the development of future offspring but also studies its acoustic properties. Having found a suitable cavity, the male enters it and makes a series of test sounds of different frequencies, studying the effect of resonance. A cavity that does not meet the conditions of resonance and therefore amplifies the sound poorly will be rejected."

Figure 1.1 It may seem to you that we are croaking, but in fact we are singing a song.

"Yes, I see, they serenade, and the vocal sacs are used to increase the volume of the sound so that as many females as possible can hear them."

"That is indeed the case. Just notice how the size of the vocal sac changes during the croaking. As the frequency of the sound changes (or the wavelength changes), the size of the vocal sac on the frog's head changes. Listen and watch carefully. Here the male is making a high-frequency sound (smaller wavelength) and the vocal sac is small in diameter at this time. But now it will change its tone toward the bass (low frequencies, and therefore longer waves) and the diameter of the sac will increase!"

"Yeah, that means it will change the diameter of the sac to meet the resonance condition: $l_i = \lambda_i n/2$. But it does it in an even more sophisticated way than happens in a violin! It doesn't need a body of complex shape, made of noble wood, dried and varnished in a special way like a Stradivarius. They simply change the size of their resonator!

"What was that again?" she interrupted. "Look! A bat attacked one of the male frogs."

"It seems the male's amorous serenade was heard by bats that feed on frogs.

"Oh no...!!!! Look! Another male frog was snatched and carried away by a bat.

"It seems that bats only attack males. And I think I know why."

"Why???"

"The fact is that bats use what's called acoustic location to hunt. They emit ultrasonic waves that bounce off various surfaces and return to a special receiver located in the bat's head. It can be said that it has a special on-board computer in its head that processes the reflected ultrasonic waves and creates a complete picture of the surrounding objects and living beings. This is why bats can fly and hunt in total darkness."

"Yes. But why do bats attack males and not females?"

Because the same vocal sac-resonators on the males' heads perfectly reflect the ultrasonic waves emitted by the bats' radars. As a result, the bat sees the singing male perfectly and chooses him for their dinner. In addition, the pulsating vocal sac is clearly visible on the bat's radar, much like the flashing light of a lighthouse seen by a ship's navigator at night. Bats even orient themselves when catching frogs by the ripples that appear on the water when they perform their "premarital serenades" and do not disappear even when the male becomes silent at the sight of a predator. The waves continue to move through the water for several seconds, creating a circular "target" around the frog, which the bats use. Although the auditory apparatus of bats is mostly tuned to ultrasonic frequencies, frog-eating bats have another peak in hearing sensitivity at a low frequency, around 5 kHz. It is in this range that the croaks of many frog species are loudest. Thus, when hunting frogs, a bat will first focus on the croaking of frogs with the aid of a low-frequency sound receiver. And then, for more precise targeting, it uses ultrasonic localization.

"It turns out that bats know physics well and use that knowledge to hunt. But it's all harsh and unfair."

"Maybe, but the concert goes on and the rest of the males keep singing. As far as I know, the frogs know how bat radar works, and have already developed some. defenses. For example, the most intelligent males sing in parts of the lake where the water surface is covered with fallen leaves. The leaves dampen the ripples of water around the singing male, depriving the bats of an additional target. As a result, when leaves appear on the water, bats are noticeably more likely to miss. There are many more interesting stories to be told about frogs and bats. But now it's late already and it's time for us to go home. Did you enjoy the concert?"

"That's a weird question. Imagine we went to a concert of some famous choir in the city and during the performance, several performers were eaten by vampires right on the stage. Would you like that?"

I wouldn't. But life goes on. Life has to go on. "The show must go on..." as the famous song goes.

Chapter 2

A Magnetic Map Encoded in Genes

I had never seen my parents nor heard them. When I first opened my eyes, I was greeted by bright sunlight and the sound of waves. Around me, my siblings were bustling about. Suddenly, as if called upon by our ancestors, we all headed toward the source of the wave sounds. It was my first and most terrifying journey in life. Fear surrounded me as my brothers and sisters were snatched up mid-run by some giant, terrifying flying ticks. These were huge seagulls. And there I was, unwanted by anyone in this dangerous world, finally reaching the sea. It was so vast, and I was so small. Where to swim in this endless sea? I needed to swim somewhere to find food and escape my cruel pursuers. As I pondered this, a clear map indicating my swimming route emerged in my consciousness. It was the Earth's magnetic field map, encoded by my parents into my consciousness. This map, hundreds of thousands of years old, seems to be passed down from generation to generation, possibly through a specific encoding in my genome. It's unusual, but the map appears in my consciousness, and I see a compass needle clearly indicating my direction of movement at every moment. It felt like I had found an old pirate map showing the way to a treasure chest buried in the ground. But my path to it—to a bounty of food—seemed very long and dangerous. Thousands of kilometers I had to swim underwater, guided only by this literally unforgettable magnetic map.

Shadowless Squids: Stories of Physics in Nature
Vitalii Zablotskii and Tatyana Polyakova
Copyright © 2025 Jenny Stanford Publishing Pte. Ltd.
ISBN 978-981-5129-43-4 (Hardcover), 978-1-003-57062-2 (eBook)
www.jennystanford.com

If you haven't guessed already, I am a tiny turtle that just hatched from an egg on the sandy shores of Java Island.

So, I embarked on a perilous journey. The unknown dangers terrified me, but my inner voice told me that even the smallest turtle could make great voyages and discover new worlds if she believes in herself and uses her knowledge of physics.

Once during my journey, I was caught in a fierce magnetic storm. The disturbances in the Earth's magnetic field reached 100 nT! And because of this, my magnetic map was covered in fine ripples, and the needle of my ultra-sensitive internal compass trembled, showing random directions. The magnetic storm lasted three days. Finally, the solar wind weakened, and the Earth's magnetic field returned to normal. My magnetic map became clear again, and the needle showed the correct route.

But what's this? What?—I screamed, looking at the map. I realized that instead of swimming east, during the storm I had swum north and ended up in the northern part of the ocean, where the waters were cool and deep. I swam past icebergs and saw large, beautiful ice floes floating in the distance. But I was quite cold, my metabolic processes slowed down even more. Realizing I could freeze here, I quickly swam south, trying to return to my original route. Along the way, I came across a small island where I decided to rest a bit.

"The law of energy conservation, use it," said my inner voice. We turtles know how to efficiently use our energy. We slow down or even stop to save energy when necessary. For this, I deeply tucked myself into my shell to protect against bad weather and save energy during rest and fell asleep.

I woke up to a terrible grating and shaking. Someone's sharp predatory teeth were trying to bite through my shell. Someone wants to eat me, I realized. It was time to use the laws of physics again—Hooke's law and the elastic properties of my shell. A turtle's shell has spring-like properties and can be stretched or compressed. I stretched hard in my shell, causing it to deform. Then I relaxed my muscles, and the shell, like a tightly compressed spring, returned to its original shape, striking the predator's teeth hard. That's how I defended myself from the predator and was free again.

Slowness: The key to health and longevity

My magnetic compass was in perfect order, the map was clearly visible, and I was full of energy and determination to continue my journey. You might think that we turtles swim slowly. But it's not true. Despite our apparent slowness, turtles are resilient and can speed up when necessary, e.g., when fleeing predators. Sea turtles are very agile and can swim up to 35 km/h using their powerful front flippers. The average turtle swims at speeds of 16 to 19 km/h and walks at speeds of 4.5 to 6.5 km/h. Usually, in folk folklore, the image of a turtle is associated with slowness. But on the Fiji Islands, on the contrary, the turtle symbolizes speed. The island's inhabitants respect this reptile for its impeccable orientation skills and swiftness displayed in the water.

Turtles are cold-blooded animals, so their life flows slower (even noticeable by their lazy movements), making them famous for their longevity. The long-lived champions among turtles are the Galapagos ones. These large reptiles, weighing more than 400 kg, live for hundreds of years. Among the registered records of the lifespan of Galapagos turtles in captivity, the maximum result is 180 years of existence. How can one not recall Cicero's rule here: To live long, live slowly.

But what physics lies behind the slowness and longevity of turtles? This question deserves special attention.

High efficiency

Firstly, turtles have a very high efficiency. The efficiency of muscles is the effectiveness with which muscles convert chemical energy into mechanical work. It can be defined as the ratio between the useful work performed by the muscles and the energy expended by them, $\eta = A/Q_1$. Scientists have calculated that the efficiency of a turtle can reach 80%. For comparison, the efficiency of an internal combustion engine, which determines the efficiency with which the engine converts energy obtained from fuel combustion into mechanical work or kinetic energy of motion, is about 25–30%, and the efficiency of a thermal engine, e.g., a steam locomotive, is only 8–10%. The low efficiency of thermal engines is related, in particular, to the emission and losses of ash. And the rate of heat release, in turn, increases with the speed of the machine's operation. For example, if you quickly compress air under the piston of a pump when inflating a bicycle tire, the air in the pump will heat up significantly. And this heat will be lost, as it will ultimately be transferred to the surrounding environment. But if you compress the air under the piston very slowly, its temperature will remain unchanged, meaning the work you do will be more spent on compressing the air and less on heating the surrounding environment. Thus, it can be said that, all else being equal, the efficiency is higher for slow processes than for fast ones. Turtles know this well and hence are in no hurry. But is reality really slow for them?

Slow metabolism and perception of time

Turtles have a slow metabolism, and we see that they are slow, but how do turtles themselves perceive what's happening around them? Interestingly, the speed of an organism's metabolism is related to its own sense of the passage of time.

The degree of speed of subjective time perception can be determined by establishing the minimum frequency of light flashes at which light is perceived by a living organism as continuous. This parameter is called the critical flicker fusion frequency. To establish the critical flicker fusion frequency in animals and humans, a special transparent drum with vertical dark stripes on the walls is used. Another drum capable of rotating at different speeds is placed outside the first drum. Both drums are brightly lit. When the second drum starts to rotate, the animal inside the first transparent drum feels as if the entire room is rotating. Accordingly, any animal starts to move in such a way as to stay upright on this moving floor. As the rotation of the second drum accelerates, at a certain frequency, the animal no longer feels that the room is rotating, and it stops trying to maintain balance. In fact, the rotation of the drum does not stop, it's just that for this animal, a specific flicker frequency has been reached when the vertical stripes on the walls have merged and become invisible. This is the critical flicker fusion frequency. Scientists have found that for humans, this frequency is equal to 60 flashes/s, while for a turtle—15. This means that, according to subjective turtle perception, time flows four times faster for them than for us. In other words, while we notice 4 flashes, a turtle sees only one. Therefore, a turtle lives in "its own time"; a turtle moves normally because, in its time, all events proceed differently: like in a slow-motion film. Therefore, a life lasting 180 years is not that long for it. Thus, their long life and high efficiency are due to their slowed metabolism. It not only slows down processes in the body but also affects the turtle's perception of reality. Only in our imagination does a turtle move slowly and live long; in its own world, these are not such record-breaking indicators. From a biological point of view, the turtle's slowed-down life is related to a lower frequency of electrical activity of its brain neurons compared to the activity of human brain neurons. But, truth be told, you've probably noticed that sometimes time drags painfully slowly for humans, and sometimes it flies by unnoticed. This is related to the varying frequencies of flashes of electrical activity of our neurons.

Life in thermodynamic equilibrium with the environment

Let's recall, in physics, in thermodynamics, there is a concept of equilibrium and non-equilibrium thermodynamic processes. Equilibrium processes are slow or quasi-static processes. These are precisely the processes, or close to them, that turtles use. As we've seen, slow (equilibrium) processes are more economical and ensure a longer lifespan for the living machine. The fundamentals of non-equilibrium and equilibrium processes were laid down by Nernst.

Nernst also possessed a magnificent sense of humor. When Ernst Rutherford visited him and saw a large pool in the yard where large fish swam, he was surprised by the host's hobby and asked why he bred fish, not, say, rabbits or chickens. To which Nernst profoundly replied—I prefer to breed animals that are in thermodynamic equilibrium with the environment. Breeding warm-blooded animals is like heating the world space at your own expense. So, the turtle also does not want to heat the world space at its own expense.

Magnetic sense

The orientation of turtles by the Earth's magnetic field during long journeys remains the subject of research. There are several hypotheses about how turtles can use the magnetic field for navigation.

Most hypotheses suggest that turtles perceive the Earth's magnetic field through magnetic nanoparticles located in their bodies. These nanoparticles are located in special cells in various parts of their bodies, such as the eyes, nose, and brain, and may help them navigate. But for orientation and navigation, a very detailed map is needed. The Earth's magnetic field has not significantly changed for many hundreds of thousands of years. Perhaps previous generations of turtles, traveling all over the globe, were able to create an accurate map of the magnetic field distribution on the surface of our planet. However, even today, creating such a map is a rather complex technical task for engineers and scientists. After all, it requires measuring the three components of the magnetic field intensity vector at all points on the surface of the globe. And the most difficult thing is how and on what to record this 3D map so that it can be passed down from generation to generation through millennia. Look at the map of the Earth's magnetic field. The lines on this map indicate the directions of the magnetic field at a given point on the surface, and the density of these lines increases with the increase in the intensity of the magnetic field. Would you be able to navigate using such a map?

Figure 2.1 Planning a long sea voyage.

Most likely, you would need some time to study it and prepare, plus a sense of magnetoreception, which humans do not have. But a turtle freely reads this map from the first minutes after birth, and it has magnetoreception.

However, despite numerous studies, the mechanism of magnetoreception in turtles is still unclear and requires further research. Do turtles indeed inherit only an old map from their parents? If so, this is an amazing phenomenon that remains a mystery to scientists.

Hears without ears, eats without teeth, and never leaves home

A few more entertaining facts about turtles, and we'll return to our adventurer.

Turtles do not have ears. Turtles do not have ear flaps and external auditory canals, but they are not deaf. Their auditory system consists of the inner and middle ear. Moreover, unlike other reptiles, their eardrums are not covered with scales but with a bony labyrinth sensitive to vibrations and low-frequency sounds, allowing them to communicate. "How can you hear without ears?" you might ask. The answer is simple. Have you noticed that when you hear your voice recorded on a tape recorder or another electronic device, it sounds somewhat different (e.g., the timbre of your voice changes) than when you hear it during your speech? This difference is explained by the fact that during your conversation, you hear your voice transmitted to the inner ear through sound waves spreading through the skull bones. And when you listen to your voice in a recording, the sound waves come into the ear through the air. Since the speed of sound waves and the laws of their dispersion are different in the air and in solid bodies, the timbre of your voice turns out to be different.

Turtles do not have teeth. They chew and grasp objects with horny jaws with sharp, almost knife-like cutting edges. Some turtles are entirely vegetarian, others are entirely carnivorous. Many are omnivorous—they eat everything they can find. For example, the green sea turtle has neither teeth nor keratin structures. It holds algae and plants in its mouth, relying on the serrated shape of the jaw.

A turtle's shell grows with the animal and, unlike the skin of lizards and snakes, is not periodically shed. Turtles cannot crawl out of their shell. Contrary to popular belief, turtles cannot leave their "home." Purely for biological reasons, they would not be able to survive without it. The shell simultaneously serves as skin and rib cage.

If you're a bit tired of physics and biology, we return to our traveler.

Life in travels and unexpected dangers

So, having fended off a predator's attack and having slept well underwater, I continued my long journey. I had to swim almost 13,000 miles: from Indonesia to the west coast of America. My mind, looking at the magnetic map, told me that it would take me about 600 days. During this time, I would grow and therefore would have to return to lay eggs at the place of my birth. To meet the deadline, I had to swim over 20 miles a day! The weather was wonderful, and I swam, thinking about eternity.

But suddenly, I felt that I could not freely swim any further. I looked at the magnetic map and saw that the compass needle was behaving very strangely: it would slowly turn in one direction, then in another. And this was not a magnetic storm; just a few minutes ago, everything was normal on the magnetic map. But these slow magnetic anomalies continued. The compass needle behaved as if someone was moving a large magnet around it. This is very dangerous, I managed to think. And suddenly, almost instantly, I was surrounded by a large school of fish, and they also could not move freely. We were all quickly pulled upward, away from the water. Squeezed by fish from all sides, I did not understand what was happening. Maybe we all got caught in a powerful whirlpool? But unfortunately, it was even worse.

We got caught in a fishing net. Within minutes, I was on the deck of a fishing seiner. "And here's a good catch for soup," said a bearded fisherman, and threw me aside from the huge pile of fish closer to the campus. The prospect of ending up in soup for the crew of fishermen did not please me, and I started looking for ways to escape. And such a plan emerged.

I slowly crawled away from the campus to a pile of some gray powder and buried myself in it headfirst. After some time, when no one was around, I crawled out of this pile and fully retracted into my shell so that I looked like a medium-sized gray cobblestone. A young sailor passed by and saw me. He picked me up, turned me in his hands, and with the words "caught a cobblestone again," threw me into the ocean. My plan worked. I was free again and could continue my journey.

But what lessons should I learn from this danger? Why did the needle of my internal compass behave strangely? This thought troubled me for several more days. I did not have a physics textbook to find the answer to this question. So, I started experimenting. I noticed that when large ships passed close to me, the needle of my compass began to deviate from the correct direction. Ah, that's it, I thought. So, the ship creates a magnetic field around itself, and it is precisely this that acts on the needle of my compass. To test this hypothesis, I approached large ships several times and watched the compass needle. Everything was confirmed, a ship is a big floating magnet. And from now on, I will stay away from ships.

Reader: Oh, exactly! I remember, in adventure and pirate novels, evil people used to place an ax under the ship's compass. Because of this, the ship would get off course and could easily be captured later.

Authors: You are absolutely right. And we hope we don't need to explain this effect further. But indeed, the question remains: who or what magnetizes ships? But you will learn about this in one of the following stories. And we return to our traveler.

So, my swim of almost 13,000 nautical miles and lasting about two years is coming to a happy end. And now it's time to summarize: to list some laws of physics and formulas that help me in travels and life.

Turtle laws

Hooke's law describes the relationship between the force F acting on a spring and its deformation, $F = -kx$, where k is the spring's stiffness coefficient, and x is the extension or compression of the spring. The minus sign indicates that the elastic force is directed toward the opposite of the spring's deformation.

Archimedes' Law. When turtles are in water, they use Archimedes' law, which states that a body submerged in a fluid experiences a buoyant force equal to the weight of the displaced fluid. Thanks to their shell structure, turtles can stay afloat on the water surface.

The efficiency of a thermal machine. The efficiency of a thermal machine is defined as the ratio of the work done by the thermal machine to the amount of heat received from the heater: $\eta = A/Q_1$. Note that even for an ideal thermal machine, $\eta < 1$.

Chapter 3

An Incident on the Hunt

Once, somewhere in the African savannah, a young man named Rustam befriended a young cheetah named Maui. They lived in a small hut because Rustam did not have a steady job and could not afford more comfortable housing. They also sometimes had trouble buying food. Therefore, Rustam and Maui were sometimes forced to hunt in the savannah.

The sight of a cheetah hunting a swift antelope is a unique opportunity to observe the graceful run of two beautiful creatures of nature, each fighting for its life. By nature, cheetahs are gentle, beautiful big cats but when they are hungry, they can chase their prey at high speeds, reaching up to 110 km/h, which is about 30 m/s, over short distances.

In their natural habitat, cheetahs hunt using their speed and maneuverability, usually leading to successful catches. A cheetah runs in a gallop and spends almost half of the chase time airborne.

This time, the day was hot and sunny. However, cheetahs are diurnal predators and are accustomed to hunting in bright sunlight. Rustam and Maui approached a herd of grazing antelopes and picked one for lunch. Maui understood without words and launched an attack. In a couple of seconds, he reached a speed of about 70 km/h,

Shadowless Squids: Stories of Physics in Nature
Vitalii Zablotskii and Tatyana Polyakova
Copyright © 2025 Jenny Stanford Publishing Pte. Ltd.
ISBN 978-981-5129-43-4 (Hardcover), 978-1-003-57062-2 (eBook)
www.jennystanford.com

but the antelope—quite the athlete herself—accelerated even more. Yet, for Maui, this was not a problem. With giant leaps reaching 8–10 m, he rapidly caught up with the antelope, which barely touched the ground with its hooves, making huge jumps and sharply changing direction, so much so that Maui, even being about half a meter away, could not catch it. Clearly, the antelope was not going to give up easily. This continued for about 2 min, and it seemed that the distance between the sprinters had shrunk to just a few dozen centimeters. Only one good jump was left. But suddenly, Maui stopped and slowly began to return. With a slightly guilty look and breathing heavily, he approached Rustam.

"What happened? Why did you stop chasing the prey?" asked Rustam. "We're probably going to stay hungry today."

"I reached the limit of my physical abilities," replied Maui.

"What physical limits? You're in great shape, and many would envy your abilities."

"When I was just a little kitten, my mother taught us about physics. She told us about the first law of thermodynamics, how heat turns into work, and work turns into heat and internal energy. I felt I was overheating, and my blood was about to 'boil' in my arteries. That's why I stopped chasing the prey. I hope you understand."

"Yes, you can't argue with the laws of physics. Tell me more about it."

"This has been known since the 19th century when James Prescott Joule discovered the law of the conversion of mechanical energy into heat, noticing that ocean water becomes warmer after a storm. Typically, if a system performs work, heat is released. Animal and human muscles also perform work, which heats them. The work of muscles is carried out through chemical reactions inside the muscles, where molecules of adenosine triphosphate (ATP) are broken down to release energy. The energy released during these reactions is partly converted into mechanical work, and the rest is converted into heat. As a result of this process, muscles heat up. In turn, heated muscles transfer heat to the blood flowing through the capillaries and then to the arterial blood. High temperatures (about 39–42°C) can cause denaturation of some blood proteins, i.e., structural changes that may lead to a loss of their functionality. At maximum running speed, a cheetah's body operates at its limit. At such speeds, a cheetah has only about 30 s to decide whether to stop or continue the chase. Indeed, at maximum running speed, the heart pumps blood at 39 °C or 40 °C directly to the brain. Another degree or two and it will overheat. Overheating the brain can instantly kill any living being. We, cheetahs, can feel the approach of critical blood temperature, and our mother taught us to immediately stop the chase in this case, no matter how close the prey might be.

"That's what I did," Maui concluded.

"I understand. I have no right to be angry with you. You did the right thing. But I'd like to know more about how long a cheetah can chase prey before its blood temperature rises above critical and it suffers from heat stroke. What affects the rate of temperature increase? Can you tell me a bit about the physics of a cheetah's chase?"

"Yes," Maui replied. "There's a lot to tell. But let's start in order."

"You already know that we, cheetahs, are the fastest land animals on our planet. A speed of 110 km/h, or to put it another way, 30 m/s, is very, very fast. Imagine, you say the short word 'one,' and in that time, a cheetah has covered 30 m. Given that an average cheetah's body length is 115–140 cm, this means it runs 20–25 lengths of its body in one second. For comparison, a passenger plane 40 m long, traveling at a speed of 850 km/h, covers only about 6 lengths of its body in one second. So, who is the real champion of speed? And can a cheetah catch up with a Ferrari? Let's figure it out.

Everyone knows that a cheetah is a predator and needs to be not only fast but also maneuverable. What provides the high maneuverability of a cheetah, i.e., its ability to quickly and sharply change the direction of its movement? Here two physical factors come into play: special claws and a long tail. A cheetah's paws have ridged pads and sharp, massive claws that cannot retract. Therefore, during running, the claws do not retract fully into the paws. Such claws are not for catching prey but for ensuring a reliable grip of its paws on the ground's surface. Unlike tigers and lions, a cheetah never catches its prey with claws but uses its teeth instead.

Cheetah claws serve as special spikes, providing a high degree of grip on the surface. It works somewhat like studded tires keeping a car on an icy road, only better. The spotted sprinter needs no more than 3–4 s to accelerate to 90 km/h. Compare this with the dry road acceleration time of 100 km/h for La Ferrari—about 3.4 s. But here's almost equality, you might say. However, the power of a Ferrari engine is 570 hp, equivalent to 419 kW. What about the power of a cheetah? Let's calculate it. On average, a cheetah weighs about 50 kg, with a significant portion of its weight being muscle mass. As we know, power equals the product of velocity and force. We know the cheetah's speed, 30 m/s, and we can find the force using Newton's second law, $F = ma$, where a is the cheetah's acceleration, which we can easily find knowing it accelerates to 90 km/h (=25 m/s) in 3 s. So, $a = 25/3 = 8.3$ (m/s^2), meaning $F = ma = 50 \times 8.3 = 417$ (N). Finally, the cheetah's power at maximum speed $P = Fv = 417$ N $\times$ 30 m/s = 12.5 kW, which is only about 17 hp. What do we see? The power of a cheetah is about 25 times less than that of a Ferrari, yet the acceleration time from zero to 100 km/h is roughly the same. And they heat up differently. While a car has a special efficient cooling system, a cheetah, on the contrary, is covered with fur, which hinders heat dissipation into the surrounding environment. Moreover, a cheetah is a midday predator, hunting at a time of day when all other carnivores are least active, i.e., in sunlight, which also does not help cool the body.

"Rustam: Are you saying that nature overlooked something here? Didn't give cheetahs a water-cooling system?"

"No, no," Maui replied. "We're fine, just we need to remember and take into account that during a long chase, brain overheating and instant death can occur. Essentially, we, cheetahs, have about 60–100 s. If we can't catch the prey within this time, we have to take a break of at least half an hour to catch our breath, cool down, and regain strength. Our mother taught us that it's better to conserve ourselves and let the prey go than to get injured or suffer a heat stroke."

"Rustam: Yes, you should listen to your mother. She wouldn't teach you anything bad."

"Maui: But I forgot to tell you about the role of the tail."

"Rustam: Of course, tell me."

The tail is an important part of a cheetah's anatomy and plays a role in its balance and maneuverability during running. The length of a cheetah's tail is about 60–80 cm. It serves to maintain balance and stability during the chase, as well as for sharp changes in the direction of movement. Let's consider an example. The body moves straight ahead at high speed. We want to change the direction of its velocity. To do this, we need to apply force, as dictated by the law of conservation of momentum: the total momentum of a body or system of bodies remains constant if no external forces act on the body, mv = const. Remember, in physics, (i) momentum is a vector quantity, and (ii) changing the direction of velocity, even while keeping its magnitude unchanged, means changing the momentum. But where does the cheetah get the force to change the direction of momentum? First, it's its special claws, which, as we've already mentioned, increase the grip of its paws on the surface, creating the force needed to change the cheetah's momentum or speed. But if a very sharp turn is needed, the force of the claws alone will not be enough. Here, its tail comes into play. If you observe the sharp turns of a cheetah while chasing prey, pay attention to the movements of its tail. Just before turning, the cheetah sharply throws its tail in the direction opposite the turn. For example, if a cheetah needs to turn sharply left, it makes a sharp movement of its tail to the right. Why? It makes the law of conservation of momentum work in its favor. Knowing that its total momentum (body + tail) in the direction perpendicular to the movement equals zero and this zero must be maintained until external forces start acting in this direction, the cheetah imparts momentum to its tail directed to the right (p_1), as a result of which the cheetah's body receives the same momentum but directed to the left ($-p_2$). Thus, the law of conservation of momentum is fulfilled ($p_1 - p_2 = 0$), and the sharp turn is successful. Then, the load again falls on its claws. This is how knowledge of the law of conservation of momentum and the ability to use the tail help in hunting. And, by the way, not only for cheetahs.

Reader: The comparison of the power of a cheetah and a Ferrari intrigued me. It turned out to be 419 kW in favor of the car and 12.5 kW for the cheetah. But it seems to me that such a comparison is somewhat limited. After all, a Ferrari is significantly heavier than a cheetah.

Authors: You are right. But in such cases, as car and airplane designers say, let's calculate a quantity called the power-to-weight ratio of the vehicle—the ratio of the power of the engine to the mass of the vehicle (a synonym for specific power). The mass of a Ferrari car is 1500 kg, and the mass of an average cheetah is 50 kg. Dividing 419 kW by 1500 kg and 12.5 kW by 50 kg, we get 0.28 W/g and 0.25 W/g for the car and the cheetah, respectively. And as you can see, there's practically equality here. Nature, in the form of a cheetah, has created a more perfect engine than man.

Reader: And here the cheetah caught up with the Ferrari!

But how did the story of Maui and Rustam's hunt end? As you might have guessed, they did not catch the antelope, and they were facing the prospect of going without both lunch and dinner. Somewhere in the evening, Rustam began to read a book on the benefits of vegetarianism with a sigh, while Maui went for a walk. You know, cheetahs usually hunt in the middle of the day, but sometimes, when they are very hungry, they make an exception and go hunting at dusk. That was the case this time. An hour later, Maui returned with prey, and soon both friends were cheerful again.

Figure 3.1 I feel like my blood is about to "boil."

An incident on the hunt: A joke

Once, two very famous Formula 1 drivers (let's call them Driver A and Driver B) went on a safari. They hunted during the day but couldn't catch anything. So, they decided to continue the next day and chose to spend the night in the jungle. They went to sleep near a large fire they had made.

Suddenly, in the middle of the night, one of the drivers was awakened by some noise. He opened his eyes and, to his horror, saw Driver B running quickly around the fire, being chased by a cheetah. "Run faster," shouted Driver A, "it's almost caught you!"

"Don't worry!" Driver B yelled back. "I'm three laps ahead!"

Reader: I think that was a fair cheetah, and it didn't dare catch Driver B until it had completed all three laps that separated him from his prey.

Authors: And I think Driver B just needed to last 2 min to escape.

Table 3.1   Physics laws used by cheetahs and the formulas readers should know

Laws	Equations	Units
$\vec{F} = m\vec{a}$ Newton's second law	Power, $P = Fv$	W, watt
$Q = \Delta U + A$ The first law of thermodynamics	Momentum, $\vec{p} = m\vec{v}$	kg·m/s
$m_1 \vec{v_1} + m_2 \vec{v_2} = \text{const.}$ The law of conservation of momentum	Power-to-weight ratio, $\varepsilon = P/m$	W/kg

Chapter 4

A Play: Conversation About Feelings

In the play.

Diogenes, a cat of unknown breed who works as a domestic philosopher and spends his free time (24/7) lying on the sofa.

Ralph, a German shepherd who serves as a household psychologist and part-time guard.

A physicist, a human who teaches physics to university students and gives free physics consultations to Ralph and Diogenes on weekends.

Diogenes announced, "Today, I've gathered you all to finally address our feelings. Philosophically speaking, feelings and sensations are given to us to enjoy the surrounding world. For instance, I can detect the scent and taste of sour cream from tens of meters away. I see perfectly in the dark and can hear the steps of a sneaking mouse from hundreds of meters away. My fur can feel the slightest touches, and I love being petted. What about you, dear Ralph?"

Ralph replied, "Well, there's not much to say. For example, in complete darkness, I can smell a hedgehog from 300–400 m away. Moreover, guided by its scent as if by a compass needle, I can run straight toward the hedgehog without error. My hearing is also fine. In complete silence, I can distinguish the steps of a human walking

Shadowless Squids: Stories of Physics in Nature
Vitalii Zablotskii and Tatyana Polyakova
Copyright © 2025 Jenny Stanford Publishing Pte. Ltd.
ISBN 978-981-5129-43-4 (Hardcover), 978-1-003-57062-2 (eBook)
www.jennystanford.com

nearly one kilometer away. I won't boast about my touch, not because I lack it, but because I don't particularly enjoy being petted by humans. You know, we shepherd dogs descended from wolves, and the wolf is a proud animal. For example, wolves never perform in circuses, unlike cats, dear Diogenes."

Figure 4.1 A scientific discussion about feelings.

Diogenes responded, "I ask you not to take this personally. We're here for a scientific-philosophical debate, not a parliamentary meeting. Let's now hear from the human about their feelings."

The physicist said, "Nature has endowed me with all five senses (sight, hearing, touch, taste, smell), though perhaps not as generously as you, esteemed colleagues. But they suffice for a normal life. Immanuel Kant, for instance, said that smell is 'the most useless' of our five primary senses. Probably why I don't detect the smell of a hedgehog even when it's underfoot. But if there's a mouse somewhere in the house, I clearly hear its presence, even from the other end of the house. My wife, upon seeing a mouse, emits such piercing sounds that it's impossible not to hear them. And Diogenes, your fur stands on end at the sound."

Diogenes quipped, "Are you joking? As chair, I could silence you."

Ralph interjected, "By the way, a mouse in the house is your oversight, esteemed chairman!"

Diogenes warned, "Dear Ralph, you're getting personal again. I'd like to remind you of potential consequences. Speaking scientifically, you might have to 'feel' my claws on your hide. But let's return to scientific facts. We all have five senses and correspondingly five sensory mechanisms that allow us to perceive the material world around us. Let's hear the physicist's explanation of how these sensors and mechanisms work."

The physicist explained, "Esteemed Mr. Diogenes, it's impossible to discuss the physics of all five senses in one seminar. While physics is more or less clear with sight, touch, and hearing, it's not so simple with smell and taste. There are many unresolved issues."

Ralph asked, "Are you having problems with smell after recovering from a COVID-19 infection?"

The physicist clarified, "No, I'm talking about unresolved issues in understanding the mechanisms underlying smell and taste. Interestingly, taste and smell are closely linked. But the mechanism of their interaction is not fully understood. For example, if a person's nostrils are pinched, they can't accurately identify even well-known food flavors, like coffee or sour cream. Once the nostrils are unblocked, taste sensations are restored. So, today, I propose we discuss what we know about the mechanism of smell perception."

Diogenes added, "Even humans can differentiate up to 10,000 smells, as I've recently read."

The physicist agreed, "I haven't counted, but that's probably true. And I won't even start on how many smells Ralph can distinguish."

Ralph proudly stated, "Yes, I can differentiate millions of smells. On a walk, I read scents like a fascinating book. It's known that we, dogs, can determine the time of day and the movement of air in a room by smell."

Diogenes remarked, "We cats determine time either by clocks or by our owner's smartphone. But what does science say about the mechanism of smell perception? Physicist, can you answer this question?"

The physicist responded ambiguously, "Yes and no, Mr. Diogenes. To put it briefly, without delving into details, the mechanisms of all five senses are very similar. Consider this: each mechanism involves five processes or stages. Take touch, e.g.: (1) the reaction of mechanosensitive receptors in the cell to external force; (2) activation of ion channels; (3) formation of a membrane action potential; (4) propagation of the electrical impulse (action potential) along nerve fibers; (5) the electrical impulse's arrival in the brain, its identification, and logical processing; (6) signal perception, image formation, and its comprehension. Similarly, with vision, the first stage involves the reaction of light-sensitive receptors, followed by the same stages 2–6. Hearing begins with sensors reacting to air pressure, then follows stages 2–6. This applies to smell and taste as well. However, problems begin with the first stage. In the cases of touch, vision, and hearing, we clearly understand the physical factors leading to receptor activation and subsequent perception stages (stages 2–6): mechanical force, light quanta, and sound waves, respectively. Yet, in the case of smell, such activation mechanisms have not been identified. Here, we observe a very subtle and barely noticeable effect on the corresponding receptor cells."

Ralph added, "Yes, yes. A single molecule of butyric acid is enough to irritate a dog's olfactory cell."

The physicist elaborated, "That's exactly the issue. A single molecule produces a very minor physical and chemical impact on any of the receptors we know."

Diogenes pointed out, "Your explanation of the problem seems vague. If I understand correctly, for stage number one, the physical impacts transferring impulse and energy to receptors are: mechanical forces for touch, light waves for vision, and sound waves for hearing. However, for smell, stage number one remains poorly understood. The rest of the perception process stages are the same for all senses and related to bioelectricity. To simplify, all stages could be divided into three categories: an 'electricity on' button, transmission of the electrical impulse through nerves, and finally, signal processing by the brain. For vision, hearing, and touch, these 'buttons' are known, but for smell, they are not. Is that correct?"

Figure 4.2 The melody of scents.

Ralph joked, "You've defined the problem as clearly as Schrödinger's cat."

Diogenes thanked him, "Appreciate the compliment. But let the physicist continue."

The physicist confirmed, "Now, the problem is clearly stated. We need to find this 'button' that, when pressed, initiates the entire perception process of a particular smell. If a nose can sense a single molecule, is the force of one molecule enough to press the button and start the electrical impulse? In 1914, Alexander Graham Bell (the inventor of the telephone) posed a question to scientists: 'Can smell be measured?' More than a hundred years have passed, but Bell's mystery remains unsolved. Yet, we now know that smell begins at the molecular level. For an olfactory signal to be perceived by a neuron, a molecule of the odorant substance binds to a special protein structure in the neuron's cell membrane, called a receptor protein. Using molecular biology methods, American scientists Linda Buck and Richard Axel discovered in 1991 that olfactory neurons in mammals contain about 1000 different types of receptor proteins (humans have fewer—about 400). For this discovery, Linda Buck and Richard Axel were awarded the Nobel Prize in Physiology or Medicine in 2004. In mice, olfactory receptors are located in the epithelium at the top of the nasal cavity, right where the skull begins. Similarly, they are positioned in other animals. When we take a deep breath, molecules pass by these receptors, which, upon being excited, send a signal to our brain. However, this only explains how the perception of smells begins. The interaction of a molecule with the receptors is not entirely clear. Humans have about 400 such olfactory receptors, while mice have about 1000."

Diogenes humorously remarked, "One doesn't need to be a great scientist to conclude that. Before being adopted by a family, I lived in a school. There I received my education: completed two grades and a corridor. But I firmly know that mice sense and differentiate smells hundreds of times better than humans."

The physicist agreed, "Yes, you're right. Humans have 2.5 times fewer receptors than mice. However, the sensitivity of a mouse's nose is hundreds of times greater than that of a human's. And here lies another mystery, probably a mathematical one. Let me try to explain."

Ralph requested, "But please, make it simpler. So that even those who completed two grades and a corridor can understand."

Diogenes playfully warned, "You always try to provoke me with words. In response, I could provoke you with my claws. But let's let the physicist continue."

The physicist proceeded, "So, the discovery of a certain number of receptors presents a mathematical problem. How can a person, having only 400 olfactory receptors, perceive around 10,000 different smells? It would make sense if one receptor detected one smell. In other words, having 400 receptors would mean you could distinguish 400 smells. Researcher John Kauer likened the theory of how smells are generated to playing a piano: A piano has only 88 keys. If each key corresponded to one smell, you would only be able to detect 88 different smells. Obviously, that's not the case. But if you consider that smells are chords, then from this piano, you could produce virtually an infinite number of smells—sounds. So, if a person has 400 such keys, then synthesizing 10,000 smells would be easy, provided you have the notes for the corresponding chords. One might say that a smell is the notes of chords or a list of molecule combinations that are registered by the epithelial receptors in the nose and relayed to special olfactory neurons, and from the neurons, the signal goes to the brain."

Ralph chimed in with interesting information, "The epithelium in both nasal cavities of a human contains approximately 10 million olfactory neurons, a rabbit has about 100 million, and I, a German shepherd, have up to 225 million. The total surface area occupied by the epithelium in both halves of an adult human's nose is small, —2–4 cm²; for a rabbit, it equals 7–10 cm²; and for dogs, it's 27–200 cm²."

The physicist thanked Ralph, "Indeed, that is the case."

Diogenes, looking disdainfully at Ralph, remarked, "You forgot to mention us—cats."

The physicist summarized the mechanism of smell perception, "Imagine an electric piano with 400 highly sensitive keys covered in a sticky substance. Air containing molecules of an odorant is blown over the keys. Some of these molecules stick to the corresponding keys of our 'piano' (bind to receptor proteins) and lightly 'press' them. Beneath the keys lie tens of thousands of olfactory neurons, which, in response to the 'pressing,' generate electrical impulses of such shape, amplitude, and duration that correspond to the given chord. The list of possible chords (smells) is formed during the organism's development and is continuously updated according to accumulated life experience. Perhaps, part of the list with the most necessary chords is recorded at a genetic level and passed down through inheritance."

Diogenes mused, "Is it all so simple? Someone inside us quietly plays the piano, and as a result, depending on the melody, we feel one smell or another. I can almost sense the melodious approach of a tasty mouse."

Ralph reflected, "It sounds simple, but much is unclear to me. For example, how is such extreme sensitivity of the keys achieved? It's known that a single olfactory receptor can 'differentiate' molecules differing in carbon chain length by just one carbon atom or molecules of the same carbon chain length but differing in the functional group."

The physicist acknowledged, "Yes, there's another unresolved issue here. After all, chemists need high-precision and expensive equipment to differentiate molecules differing in carbon chain length by just one carbon atom!"

Ralph boasted, "And a dog's nose does it for free. By the way, Diogenes, here's some useful information for you: scientists found nerve cells in mice brains that awaken the animals upon the approach of a predator."

Diogenes thanked him, "Appreciated, I'll keep that in mind."

The physicist shared, "Recent experimental research on the properties of olfactory receptor proteins has enabled the creation of a structural model of the olfactory protein at the molecular level. This olfactory protein appears as a spiral molecule consisting of 300–350 amino acids, crossing the bilayer lipid membrane of a cilium 7 times."

Diogenes admitted, "Even I find that hard to envision."

Ralph interjected, "I read somewhere that molecules of enantiomers smell differently. Enantiomers are substances whose molecules can exist in two forms, mirroring each other as an object and its incompatible mirror image. For example, the natural substance limonene can exist in two forms: left-handed and right-handed. Despite their molecules being as similar as the right and left hand, right-handed limonene smells like orange, while left-handed limonene smells like lemon. Can you smell the difference, esteemed Diogenes?"

Diogenes remained silent....

The physicist clarified, "It was recently proven that the widely cited example of enantiomers of limonene having different smells, which wandered from textbook to textbook, is nothing more than a myth. Organic chemistry textbooks need to be rewritten. However, on the other hand, the olfactory protein is a spiral molecule, i.e., it possesses a certain chirality. This means that the enantiomers of a fragrant substance should react differently with chiral protein receptors of smell."

Diogenes, curious, asked, "You used the word chirality. What is that?"

The physicist explained, "What is chirality? Let me explain simply. The word comes from the ancient Greek Χέρι—'hand.' Just as the left and right palms cannot be superimposed no matter how you turn them, molecules of certain chemical substances cannot be superimposed on each other through any rotations or overlays. In other words, a hand and its reflection in the mirror have different chiralities or represent different enantiomers: left and right. Proteins, amino acids, and almost all organic substances in living nature exist in only one form of enantiomer."

Ralph followed up, "So, do molecules of enantiomers smell differently after all?"

The physicist mentioned, "Professor David MacMillan, a Nobel Prize laureate in 2021 (jointly with Benjamin List) for the development of asymmetric organocatalysis, cited in his Nobel lecture an example of enantiomers that smell differently, not limonene, but carvone: (R)-carvone smells like mint, while (S)-carvone smells like caraway. Note, scientists need sophisticated instruments to distinguish these two substances, but Ralph could sense the difference with his nose."

Ralph quipped, "Well, when you provide me with these substances, I'll try to differentiate them by smell."

Diogenes mused, "It seems we've delved into the complex scientific thicket of chirality. From what I understand, molecules of the same substance, here and in the mirror world, would smell differently."

The physicist confirmed, "Exactly."

Diogenes objected, "I think not! I venture into the mirror world nearly every day and haven't noticed anything like that!"

Ralph humorously remarked, "Dear Diogenes, hiding in the wardrobe behind the mirror doesn't mean you're actually in the wonderland of the mirror world."

Diogenes concluded, "The discussion is over for today. I hope everyone understood everything."

Chapter 5

The Magic of Magnetism

Magnetism and magic are closely related both in sound and meaning. Yet, let's immediately distinguish one from the other. Magnetism is the science of the magnetic properties of matter and magnetic fields, stretching from atomic to astronomical scales: from angstroms to hundreds of millions of light-years. Magic is the sorcery we cannot understand or explain. It's fascinating that magnetism also contains a bit of magic, meaning some magnetic phenomena, especially their manifestations in living nature, remain inexplicable at our current stage of scientific development. This refers primarily to magnetoreception—the ability of certain living organisms to sense magnetic fields.

But first, a few well-known or lesser-known facts about magnetic fields and magnetic materials. In Chinese, the word "magnet" literally means "loving stone." There are recollections of Einstein being astounded at four years old by a magnet acting through wooden planks. What we now know about magnetic fields should impress you far more than magnetic force transmitted through planks. However, there are phenomena and facts related to magnetic fields that contemporary science cannot yet explain. For the most curious readers, this should act as a magnet, drawing them toward future scientific discoveries.

Shadowless Squids: Stories of Physics in Nature
Vitalii Zablotskii and Tatyana Polyakova
Copyright © 2025 Jenny Stanford Publishing Pte. Ltd.
ISBN 978-981-5129-43-4 (Hardcover), 978-1-003-57062-2 (eBook)
www.jennystanford.com

Where lies the magic of science? It lies in the fact that initially, we understand nothing and cannot explain the observed natural phenomena. Later, as we study, we begin to understand some aspects. But the more we understand, the more new questions arise, and the more global the unresolved problems become. So, what do we already understand about magnetism, and what do we not?

We know and understand how magnetic fields are created by permanent magnets and electric currents. We know how a magnetic field acts on a current-carrying conductor and moving charges. Why and how a magnetic field turns a magnetic needle (magnetic dipole) and rotates the plane of light polarization. Why functioning ion channels of cellular membranes generate weak magnetic fields, tens of pT. We know that a weak magnetic field arises in the stems of some plants, e.g., in the carnivorous Venus flytrap that feeds on flies carelessly lingering in its toothy "jaws." Functioning neurons of the brain generate weak, but measurable, magnetic fields. Humans create a small magnetic field around themselves. We know that rapidly spinning neutron stars (magnetars) create enormous magnetic fields around themselves, reaching 10^{10} T. We understand how a comparatively weak magnetic field rotates the rotor of an electric motor in a car, thus moving an electric vehicle. How permanent magnets and magnetically controlled micro-robots are used for surgical operations. Large superconducting magnets, generating magnetic fields ranging from 3 to 10 T, are used in medicine for diagnostics. These are MRI scanners. We know what magnetic levitation is and use it in maglev trains that move without touching the rails. With the help of a magnetic field, we can hold plasma heated to hundreds of millions of degrees so that it does not touch the walls of the toroidal chamber (tokamak—a device for conducting thermonuclear fusion reaction). Magnetic recording of information is used in computers. We can create new magnetic materials with the properties we need. It is the Earth's magnetic field that helps gravity keep our planet's atmosphere. For example, Mars has no magnetic field because its core has cooled, and its atmosphere has been almost entirely blown away by solar wind. We know that the magnetic field protects life on our planet from solar radiation: the geomagnetic field deflects solar wind streams from Earth into space. Every second, 1.5 million tons of solar wind, mainly consisting of ions, are ejected from the Sun into space at hundreds of kilometers per second. If Earth had no magnetic field, much of this would fall on our heads. We know much more than listed above. But it's far more interesting to learn what we still don't know.

We still don't understand or know: Does the Dirac monopole—a hypothetical particle representing a single-pole magnet (unlike the well-known magnet with two poles), theoretically predicted by P. Dirac in 1931, exist? Why does an electron have a spin and a corresponding magnetic moment? We know practically nothing about the effect of the magnetic field on the properties of ordinary water. Is the emergence of life possible without the Earth's magnetic field? How does the magnetic field select the chirality of biological molecules (you can read about chirality and its role in living nature in one of the following stories)? Why can a magnetic field applied to a certain part of the brain change left and right in a person's consciousness? For example, under the influence of a magnetic field, right-handed people can temporarily become left-handed or vice versa. Why does the magnetic field distinguish left from right (remember the left-hand rule for determining the force acting from the magnetic field on a current-carrying conductor)? We do not know how a strong magnetic field can have an antidepressant effect on mice and increase the level of the happiness hormone (oxytocin) in the brain. We do not know the cause and mechanism of solar magnetic storms and cannot predict them. We do not understand how the Earth's weak magnetic field and its slight disturbances during magnetic storms affect people's well-being and health. How living organisms, including humans, can feel not only the magnitude of the magnetic field but also its direction?

Magnetoreception: The sixth sense of living organisms and the seventh sense of humans

It is known that many living organisms have magnetoreception—the ability to sense weak magnetic fields. Despite centuries of study, its mechanism remains a mysterious phenomenon.

Diogenes: Stop. Why is magnetoreception the sixth sense of animals but the seventh for humans?

Authors: Both humans and animals have 5 senses: touch, smell, sight, hearing, and taste. But it is believed that humans have a sixth sense—intuition. Therefore, in the list of human senses, magnetoreception occupies the seventh place.

Diogenes: You're offending us.... We cats also possess intuition. And somehow, I will prove it to you.

Authors: Apologies, esteemed Diogenes. We are merely presenting the commonly accepted view and do not deny your capabilities. So, magnetoreception is the most mysterious of our senses.

Naive reader: What's so mysterious here? As far as I'm concerned, it's been quite clear since ancient times. The compass needle shows us the direction of our planet's magnetic field. So, look for the compass needle in a living organism.

Insightful reader: A typical compass needle contains a lot of ferromagnetic material, such as iron. Therefore, it can be rotated by the Earth's relatively weak magnetic field, more precisely, by its horizontal component. I don't think macroscopic needles made of iron can be found inside living organisms.

Authors: Indeed, there are no macroscopic magnetic needles there. But magnetic micro-compass needles exist in the form of chains of magnetic nanoparticles of iron compounds (Fe_3O_4 or Fe_3S_4) in so-called magnetotactic bacteria, first discovered in 1975 by Richard Blakemore.

These bacteria have learned to synthesize iron nanocrystals within themselves, forming magnetic chains—microcompass needles. Using such an internal microcompass, bacteria can orient themselves by the Earth's magnetic field just as tourists navigate unfamiliar terrain with a compass.

Ralph: But what unfamiliar terrain are we talking about if you're referring to bacteria?

Authors: These bacteria live in murky water, even in ordinary puddles. So, the environment in which they dwell is poorly transparent to sunlight, and bacteria lack sensory organs. Note that, like all living organisms, bacteria need oxygen to breathe. Now, dear Ralph, imagine you are in a homogeneous opaque liquid medium, where gravity is compensated by the buoyant force of water, there's practically no light, no sounds, and no smells. And you want to emerge from this liquid to breathe fresh air. Note, the concentration of oxygen dissolved in water increases as you rise to the surface. But how to know where is up and where is down?

Ralph: Look at the compass, reacting to the vertical component of the geomagnetic field. Or better yet, feel where the compass needle inside you turns.

Authors: Correct, you need to use an internal compass. Therefore, magnetotactic bacteria have learned to synthesize magnetic needles—chains of magnetic nanoparticles inside themselves. It is important to emphasize that for their compass, bacteria do not take ready-made magnetic nanoparticles from the external environment but synthesize them themselves at their internal microfactory. To do this, they must have a whole set of proteins capable of changing the valency of iron ions and controlling the growth of iron crystals. These are advanced nanotechnologies used by microorganisms. Bacteria have not only learned to produce perfect magnetic nanocrystals but also recorded the method of their synthesis in their genome.

Diogenes: Do you want to say that the magnetoreception of birds, marine organisms, and other animals was inherited, i.e., passed along with the genes from the simplest bacteria?

Authors: Why not? It is known that living organisms evolve from the simplest microorganisms to more complex ones. So, the transfer of genes responsible for the synthesis of magnetic nanoparticles from bacteria to more complex organisms is quite possible. Inside many living organisms, scientists have found a magnetic nanoneedle that can react to the geomagnetic field. Such a needle turned out to be the complex of proteins Cry and MagR with iron. Together, these proteins indeed form something like a needle that can track not only the direction but also the intensity of the magnetic field. However, both of these proteins are found in many animals, including mammals. Against this backdrop, the human magnetic sense doesn't seem so incredible.

In 2016, American geophysicist Joe Kirschvink reported that he and his colleagues had found a magnetic sense in humans: the brain's electroencephalogram of volunteers participating in the experiment showed how their brains reacted to changes in the magnetic field around their heads. Experiments were conducted in an underground laboratory, where the geomagnetic field was shielded (so-called Faraday cage). A small uniform magnetic field was created using a system of three mutually perpendicular coils with current. This arrangement of coils allowed controlling the direction of the magnetic field in the laboratory. The brain's reaction to the magnetic field was recorded using electroencephalography (EEG) by observing alpha waves—electrical impulses in the brain with an average amplitude of 30–70 μV in the frequency band from 8–14 Hz.

Don't be alarmed, these alpha waves are not similar to sea waves, and don't look for them on the surface of your head. Alpha waves are just propagating oscillations of electrical activity in the brain. After all, you probably know that all our thoughts consist of a set of diverse electrical impulses spreading through the neurons of the brain. The visible change in alpha waves on the EEG was supposed to indicate new brain activity centers (new human sensations), occurring with a change in the direction of the magnetic field in the laboratory.

Several tests were performed on volunteers: in some cases, a magnetic field slowly rotated around the head, approximately equal to the Earth's field, in others, only the natural geomagnetic field affected the person. Tests alternated randomly, so neither the experimenter nor the subject knew which test was happening at the moment. It turned out that when the magnetic field rotates counterclockwise (as if the subject looked at an imaginary clock face, which is placed horizontally), the alpha waves sharply weaken, and the same happens when the magnetic field tilts downward (corresponds to looking up). Interestingly, small doses of caffeine (one cup of coffee) also cause a decrease in human brain alpha wave activity. The decrease in alpha wave activity is observed in those parts of the brain that are activated at the moment.

Thus, the rotation of the magnetic field counterclockwise and caffeine cause similar effects of decreasing alpha wave activity, thereby opening certain brain areas for storing useful information or making decisions. Why alpha waves react precisely to such changes in the direction of the magnetic field and do not respond to the opposite directions, is still absolutely unclear. But, undoubtedly, these experiments, testify to the existence of a magnetic compass in humans. Perhaps, we have an internal magnetic compass, but evolution has not taught us to use it. Apparently, evolution decided otherwise: it did not give humans time and opportunity to learn to use their natural internal compass, but instead gave them intellect, which allowed humans to first make a compass in the form of a magnetic needle on a thin needle, and later create sophisticated navigation systems, e.g., GPS.

Diogenes: Are you saying that I also have a hidden magnetic compass inside me, but I don't know how to use it? According to you, it turns out that, e.g., dogs and cows also have a magnetic compass inside themselves.

Authors: Yes, perhaps that's the case. There is a large number of scientific publications that claim that cows, deer, and dogs possess magnetoreception and can orient themselves by the magnetic field when needed.

Ralph: I wouldn't believe you about cows. They simply don't need a compass. Cows, fortunately, do not fly, which means they graze not far from home and therefore easily find their way back home without using their internal quantum compass. I also want to say that I found it very interesting to learn that coffee and the rotation of the magnetic field counterclockwise have similar effects on the human brain. But on the other hand, it is known that caffeine improves human cognitive abilities, which any student drinking more than one cup of coffee before exams can confirm. From this, I conclude that the magnetic field can also improve human cognitive abilities. Is this really the case?

Authors: Dear Ralph, congratulations! You've made an absolutely correct conclusion. There are dozens of scientific publications experimentally confirming that the magnetic field improves the cognitive abilities of mice. But you are the first to see the analogy between the action of caffeine and the magnetic field.

The inversion of the Earth's magnetic poles could have been the cause of the mass extinction of living organisms

Authors: The Earth's magnetic field and life on our planet are very closely related. And this connection is even stronger than we might imagine. For example, a change in the magnetic field that occurred 42,000 years ago may have contributed to the mass extinction of animals. During the period of the Earth's magnetic poles inversion, the geomagnetic field was only about 28% stronger than today. But during this transitional period, occurring approximately from 42,300 to 41,600 years ago, the field's intensity was reduced to about 6% of the current level.

Researchers compared periods of geomagnetic field weakening with previous ice core records, which reflected changes in solar activity, and concluded that solar activity was at a minimum at that time. Such a simultaneous combination of a weak magnetic field and reduced solar activity "created a perfect storm" of climate and initiated large-scale environmental changes, putting severe strain on the populations of all living organisms. Thus, it can be argued that the weakening of the Earth's magnetic field correlates with cascades of ecological crises on our planet. For example, one such cluster of Australian megafauna extinctions, including the disappearance of Diprotodon and the giant kangaroo (*Procoptodon goliah*), occurred about 42,000 years ago. The thickness of the ozone layer, which protects us from solar ultraviolet radiation, is important for the animal world. During the inversion of the magnetic poles, the thickness of the ozone layer decreases. This is indirectly confirmed by handprints covered with red ochre, dated almost 42,000 years ago, found on the walls of the El Castillo cave in Spain. Since red ochre was used as sunscreen in ancient times, a large number of handprints covered with red ochre left by people on cave paintings possibly indicate an excessive flow of ultraviolet radiation, forcing people to seek shelter from the intense sun in caves and use sunscreen—ochre.

Ralph: But that was a long time ago. What's happening with the Earth's magnetic field now?

Authors: Data obtained from European Space Agency satellites show that the Earth's magnetic field is mysteriously weakening. Scientists believe that such rapid field weakening is a harbinger of an imminent inversion of the magnetic poles.

Diogenes: So, scientists predict the imminent end of the world? A magnetic end of the world is coming, and we will all perish! Right?

Authors: There's no need to panic in advance. First, the word "soon" (inversion will occur) should be considered on a geological time scale. It could be tens or even hundreds of thousands of years. During this time, humanity will discover many new laws and physical phenomena that will allow scientists to find effective protective measures against disturbances in the geomagnetic field. Second, scientists do not know how long the process of magnetic poles inversion will last. Will it be a quick process (days or months) or will it stretch over a thousand years? In the latter case, people and animals have a chance to adapt both to the new magnitude and the new direction of the Earth's magnetic field.

Diogenes: So, if the inversion (reversal) goes slowly, we just have to repaint the compass needles: paint the blue end of the needle red and vice versa.

Figure 5.1   With a compass to black holes.

Authors: Yes. This is the simplest option. But don't forget, during the inversion of the magnetic poles, a situation may arise when the Earth simultaneously has two southern magnetic poles and two northern poles. For example, the magnetic fields of Uranus and Neptune, unlike all other planets in the solar system, are not dipolar but quadrupolar, meaning they have 2 northern and 2 southern poles each. The magnetic field between the four poles of the magnets (quadrupole field) will change very sharply both in direction and magnitude when moving from one point in space to another. Therefore, your compass won't help you either on Uranus or Neptune.

But, what do we see? Diogenes is very sad and clearly upset. What's wrong with you, Diogenes?

Diogenes: I feel sorry for the birds. How do they fly miserably on Uranus or Neptune? And what will happen to their magnetoreception on Earth during the inversion of poles? My intuition tells me that birds will simply get confused by their quantum compass readings and won't find their way to their habitats. Their internal compass will go crazy, torn between four magnetic poles.

Authors: Of course, you're joking, Mr. Diogenes, when talking about the unhappy migratory birds on Uranus and Neptune. But the question is very good: what would bird magnetoreception look like in a quadrupolar magnetic field? Maybe evolution will give birds and other animals two internal quantum compasses. But no, that won't help. Let's leave this question to future generations of scientists.

In summary, it can be said that the magnetic field literally surrounds us, extending from atomic scales (the magnetic field of nuclei and electrons moving in an atom) to astronomical ones. The magnetic field of galaxies extends for hundreds of millions of light-years. Even black holes have been found to have a magnetic field. If we consider the scale of magnetic fields' extension in terms of life on Earth, the known forms of life exist on lengths from molecular (10^{-8} cm) to tens of meters, which is undoubtedly the zone of magnetic fields' influence. But could there exist forms of life on astronomical lengths? For instance, imagine a creature several light-years long, flying from one galaxy to another using a sail and light pressure as the propelling force and feeding on pure energy. We don't know if there are any forms of life not tied to a specific planet or a star's planetary system. Undoubtedly, if such trans-galactic forms of life exist, then the magnetic field would significantly influence them. For instance, to navigate the endless and dark space using intergalactic magnetic fields, our gigantic cosmic sailer would need a sense of magnetoreception. For this, it would require an internal magnetic compass not of microscopic but of astronomical size. And the magnetic fields of black holes would serve as excellent beacons for these cosmic wanderers.

Reader: I have a question about the magnetic field of a black hole. Where does it come from? Everywhere it's written that a black hole swallows everything like a hungry cosmic shark and nothing comes out of it. How does the magnetic field manage to escape from the jaws of a black hole?

Authors: You've touched on one of the global questions of astrophysics. Hypothetically, there are several mechanisms. For example, black holes born from the collapse of magnetized stars are born with a magnetic field penetrating the event horizon. In astrophysics, the event horizon is the boundary beyond which events cannot affect an observer. In other words, the event horizon is the boundary of a black hole or the boundary of the no-return zone for a body approaching a black hole.

So, in the mechanism mentioned above, it is already assumed that the magnetic field escapes the grasp of the black hole's gravity (or penetrates the event horizon) immediately upon its birth. According to another hypothesis, a black hole can acquire its own magnetic field as a result of merging with a magnetized neutron star (magnetar). Because of this, the black hole acquires hairs in the form of magnetic field lines. But the magnetic field can evaporate and leave the black hole "bald."

Diogenes: "Bald black holes!" Who could have dreamed of such a thing? We started with bacteria and birds, and we end up with bald cosmic magnetic monsters living on the event horizon.

Ralph: I agree. I like hairy black holes more. Don't forget that matter falls into a black hole at tremendous speeds. And directly near the black hole, the matter consists of plasma streams, i.e., fast streams of charged particles. And as is known, a magnetic field arises around moving charges. So, black holes are as if covered with magnetic hairs—long fur.

Authors: Yes, along with the magnetic field, we have already traveled both in time and space. We hope that during this journey, readers saw many mysterious manifestations of magnetism and felt the need for a deep study of natural sciences to understand and explain not only these phenomena but also those yet to be discovered.

Chapter 6

Mysteries of Bats

We became curious about what schoolchildren know about bats. So, we visited a school and asked the students 10 simple questions.

Can a bat see a cat in complete darkness?

Is a bat a mammal or a bird?

Can a bat fly in a confined space, such as a cave, if its eyes are closed?

And if a bat's mouth is closed, can it navigate and fly in a confined space?

Can a bat navigate in space with its ears closed?

Why does a bat sometimes crash into a person's head?

What benefits do bats bring to humans?

What laws of physics do bats know and use?

Can a bat perform an ultrasound?

Can a flock of bats eat 200 tons of mosquitoes in one night?

Thirty students participated in the contest. Below, we present the most interesting incorrect and partially correct answers they provided.

Answers to question 1 included: Yes, if the cat starts sparking. Yes, if the cat is white with glowing eyes. Yes, if it's a cat with wings (our Diogenes smiles). No, if it's Schrödinger's cat.

Shadowless Squids: Stories of Physics in Nature
Vitalii Zablotskii and Tatyana Polyakova
Copyright © 2025 Jenny Stanford Publishing Pte. Ltd.
ISBN 978-981-5129-43-4 (Hardcover), 978-1-003-57062-2 (eBook)
www.jennystanford.com

Answers to question 2: It's not a mouse, not a bird, and not a mammal; it's a vampire. Bats drink blood, not milk.

Answers to question 3: Yes, but not for long, until it collides with an obstacle, as it sees nothing.

Answers to question 4: Yes, with its mouth closed, the bat will fly even better, as it will not be distracted by talking and thus be more attentive.

Answers to question 5: There would be no difference in flights with ears open or closed.

Answers to question 6: A bat crashes into a person's head to scare them more. To drink blood directly from the head.

Answers to question 7: Bats are beneficial to humans because they feed on insects such as mosquitoes and other garden and field pests, helping to reduce the populations of harmful insects. Some bat species also act as pollinators. They may feed on nectar and pollen, transferring them from flower to flower, facilitating pollination, and improving the yield of some plants.

Answers to question 8: Bats were not taught physics. They know the laws of gravity and use them to sleep upside down.

Answers to question 9: No, it cannot. It does not have the special equipment for that.

Answers to question 10: No, it cannot. There aren't that many mosquitoes (about 5 railway wagons) anywhere.

Please note that although some of the answers above contain correct information, they cannot be considered fully correct or comprehensive.

Reader: And when will the correct answers be given?

Authors: We will not provide correct answers. If you read this story to the end, you will easily answer these and other questions yourself.

So, let's begin.

Bats are mammals, not birds. They are small nocturnal animals belonging to the order Chiroptera and have specialized anatomical adaptations for flight, such as wings made of a skin membrane stretched between the fingers. Bats are the only mammals to have mastered true, flapping flight. Their forelimbs have transformed into wings by elongating the bones of the forearm and hand, which serve as the framework for a thin, elastic skin flying membrane stretched between them, the sides of the body, and the hind limbs.

Physics: Bat echolocation

With exceptionally keen hearing, bats are the animal kingdom's champions at receiving sound waves. The range of sound waves they can perceive is very broad, from 12 to 190,000 Hz (for comparison, humans can hear sounds in the range from 20 to 20,000 Hz). Therefore, for them, hearing is more important than sight. They navigate perfectly in complete darkness without bumping into obstacles. Moreover, having poor vision in the dark but excellent hearing, a bat can detect and catch small insects while flying.

Scientists in the 18th century established that complete darkness does not hinder bats' orientation and flight. When bats' eyes were covered, it changed nothing in their behavior; they still hunted insects excellently. But if their ears were covered, they began to blindly and chaotically bump into obstacles in their path and could not hunt. Thus, it can be said that bats "see" with their ears. Scientists, using a sound detector capable of capturing sounds across a wide frequency range, discovered that bats emit sounds beyond the threshold of human hearing. It turned out that most of our contest participants did not know that for orientation in space, bats need not only their ears but also their mouth, more precisely their vocal cords, and even more precisely, a generator of sound waves.

A vast number of bats can fly simultaneously in a dark cave without colliding with obstacles or each other. For example, Bracken Bat Cave in Texas is considered the largest bat colony in the world. More than 20 million bats live in this cave, more than the population of Mumbai, an Indian city that is one of the most populous cities in the world. When the bats leave the cave, their group is so large that it is detected on aviation radars as a huge pre-storm cloud, which eats about 200 tons of bugs in just one night. How do bats in this cloud avoid colliding with each other? It turns out that a slight change in the frequency of the ultrasonic signal (by 3–6 kHz compared to the general broadcasting band, which was 75–80 kHz wide) allows bats to clearly determine from where and from what obstacle it was reflected. Information about the reflected sounds is processed in the bat's brain, which thus becomes capable of avoiding both moving and stationary obstacles in its path.

How does a bat "see," i.e., orient itself in complete darkness? As we already mentioned, a bat emits directed ultrasonic waves, then receives and analyzes the waves reflected from various objects. In other words, the bat feels objects with its signals—ultrasonic waves. The number of signals changes depending on the distance between the bat and the object. If the distance to the object is small, the bat sends signals more frequently. For example, at a distance of 20 m to the object, the bat sends 5–8 signals, and at a distance of 1 meter, about 60 signals. This principle is also the basis for the operation of ship echolocators, which measure the depth of the sea floor. Knowing the speed of sound in the air (330 m/s) and the time between the emission and reception of the signal, the bat calculates the distance to the object. Moreover, it does this very quickly, as the bat's flight speed is quite high. If there is a delay in calculations, a collision is inevitable.

When a bat is in flight, its mouth and nose act as an emitting antenna for sound waves, a kind of sound "spotlight." It "illuminates" the path with a narrow sound beam. The bat's huge pinnae are directed in the same direction and catch the reflected ultrasound. And if the signal is not reflected from anywhere, the ears hear nothing. This means that there is empty space ahead, and the path is clear. Of course, if objects that absorb ultrasonic waves well are in the bat's path, there will be no reflected signal, and the bat may simply crash into such an object. For example, into a human head with thick hair or even a fluffy cat.

Diogenes: Please, leave me alone.

Authors: Otherwise, such ultrasonic location works excellently. A bat can distinguish the echo from a stationary obstacle and the echo from a moving object.

Ralph: This means that bats know the Doppler effect—the change in the frequency of a wave perceived by an observer due to the motion of the wave source relative to the observer. Listen to the sound of a motorcycle approaching you, and then to the sound of a motorcycle moving away. You should notice that when the motorcycle approaches, the sound frequency it emits is higher than the frequency you hear when it moves away.

Authors: A bat can hear the faint echo of a flying mosquito against the much stronger echo from the ground, trees. If the path is clear, the bat flies straight; if there is an obstacle, the bat will hear the echo and turn aside. The maximum distance at which a bat feels an obstacle is about 25 m. Thus, with the help of an echolocator, bats not only orient themselves in space but also hunt for mosquitoes, moths, beetles, and other nocturnal insects.

Reader: Why don't these animals go deaf from the large amount of ultrasonic noise they emit?

Authors: Nature has provided for this as well, equipping bats with special muscles that protect their pinnae from unwanted sounds.

Reader: Can we say that the bat's generator and receiver of ultrasonic waves are constructed similarly to a medical ultrasound device? Essentially, during an ultrasound procedure, a doctor "feels" a sick organ using ultrasonic waves. But I think if we taught bats to perform ultrasound on humans, their diagnostic accuracy would be higher than that of modern medical devices. After all, bats can detect a mosquito several meters away.

Authors: Yes, you can. But everything is the other way around. The first ultrasound devices appeared in 1949, while bats have been using ultrasonic location for hundreds of thousands of years. So, it's better to say that ultrasound devices use the method of ultrasonic location developed by evolution specifically for bats.

Remember, we started this story with questions. To the question: Is a bat a mammal or a bird? We received an unusual answer. It's not a mouse, not a bird, and not a mammal; it's a vampire. So, let's talk about vampires.

A bit of vampirology: Physics and chemistry at the service of vampires

There are about 1300 species of bats in the world, but only three of them have learned to drink the blood of warm-blooded animals. Evolution has given them a special thermosensitive sensor (organ) that allows them to find the optimal place for a bite—a place where a blood vessel approaches the skin surface. Don't be afraid, vampire bats bite almost painlessly, as their saliva contains substances with an anesthetic effect that prevent blood clotting. Similar substances are also found in the arsenal of blood-sucking mosquitoes.

So how do vampires draw blood? Much like how a nurse takes blood for analysis at a clinic. First, the vampire prepares the victim's skin area for the operation, plucking hairs or feathers and licking the bare spot for disinfection and anesthesia. Then, it plunges its sharp teeth into the bloodstream and begins to drink blood. Proteins in the bat's saliva prevent blood vessels from closing (increasing blood pressure) and blood from clotting, so bleeding can continue for several hours. One blood-drinking session can last from 20–30 min to an hour. During this time, the size of the vampire can double thanks to its stomach's ability to stretch. After such a hearty meal, the vampire has trouble taking off, like an overloaded vertical take-off aircraft. Vampires usually suck the blood of domestic mammals, most often cattle and horses, but they do not ignore wild animals.

Reader: These aren't vampire bats; they're certified pharmacists! Yes, I see that studying them is undoubtedly beneficial for pharmacology and ultrasound diagnostics.

And again, physics

Diogenes: I would like to add about the verbal abilities of ordinary house mice. I'm sure you've heard something similar to chirping or whistling from mice. Scientists have studied the verbal signals of house mice and found that males attract females exclusively with ultrasound. And "ladies" use the highest notes when communicating with "gentlemen." Although among themselves, females can communicate normally, but with males—only in raised tones.

Reader: I have a question for Diogenes, as an expert on mice.

Diogenes: I'm listening.

Reader: Why do house mice have such long tails? It seems that navigating through holes would be more convenient with a short tail or no tail at all.

Diogenes: Allow me to answer based on the laws of physics and evolution. As you know, mice don't like to work but prefer to steal food from humans. But thieves often face chases. After all, cats have lived with humans for a long time and protect their food supplies from mice. And now, briefly, the physics of the chase. Imagine a mouse running away from a cat. If it runs straight, the cat can easily catch it. But if the mouse makes zigzags, i.e., abruptly changes the direction of its movement, its pursuer—the cat—must make the same zigzags to catch it. But it's harder for the cat because it has a larger mass, and therefore, greater inertia. It's also not easy for the mouse to abruptly change its direction of movement. But its tail helps in this. And the longer the tail, the more sharply the mouse can change its direction of movement. In this case, the mouse sharply turns its tail to one side, causing its body to turn in the opposite direction. Here, the law of conservation of momentum in the direction perpendicular to the mouse's trajectory of movement works. So, you guessed it, if the tail is short, its mass is small, and therefore, the change in direction of momentum it can give to the body of the running mouse is also small. The conclusion, based on the law of physics, is unambiguous—a mouse with a short tail will be caught by a cat or a human with a higher probability than a mouse with a long tail.

And now let's turn to the laws of evolution. Do they confirm the conclusion made from the laws of physics? Biologists and archaeologists have conducted research on the fossils of house mice. And they found that the population of the first domestic rodents was quite small. At that time, long-tailed rodents competed for provisions with short-tailed ones, and the latter clearly lost in maneuverability. Human settlements provided mice with mountains of provisions. The house mouse with a long tail, thanks to its tail, managed to escape from cats and humans, while the mouse with a short tail did not. Therefore, over time, mice with short tails disappeared. That's how evolution solved this issue.

Reader: Thank you. I understand everything. A long tail is a mouse's savior.

Ralph: But bats don't have tails. How do they maneuver and escape from enemies?

Diogenes: Bats have few natural predators. And since cats, unfortunately, can't fly, bats have no worthy opponents in the air. Of course, some predatory birds, such as owls, hawks, and falcons, can attack bats in the air, and snakes on the ground. But these are rare cases. Therefore, nature, knowing the aerodynamics of mouse flight, decided not to give it such a tail as a house mouse. For maneuverability, a bat's unique hands—wings—are quite sufficient.

Authors: Bats are the only winged creatures capable of very long and high-speed horizontal flights. Female bats (*Tadarida brasiliensis*) weighing 11–12 g can reach speeds of up to 160 km/h. For comparison, the maximum speed of horizontal flight for a common swift is about 120 km/h.

The main mysteries of bats

Bats live a very long time!

On average, a 7 g individual can live up to 43 years, which, if converted to human lifespan, would be approximately 1000 years of human life. No other animal in the world of similar size and metabolic rate lives as long. Why do we find this fact so mysterious and mysterious? The fact is that there is some correlation between the lifespan of animals and their mass, although it is not absolute and unequivocal. The general trend is that larger animals usually have a longer life than smaller ones. This is hypothetically explained by the fact that larger animals have more complex organisms, a longer development process, and a lower metabolic rate. This may contribute to less susceptibility to diseases and a longer life. And here, the bat is an exception that scientists cannot explain.

Bats are never fat!

Here are two shocking facts that show how high the metabolic rate is in bats. A small bat can swallow six hundred mosquitoes in 60 min. For a human, such a portion of food would be equivalent to twenty pizzas eaten in an hour. Bats digest food very quickly, e.g., bananas and berries eaten by them are completely absorbed by the body within 20 min. From a physics point of view, this is quite justified, as the flight of a bat, having a poor aerodynamic body shape, requires not only high energy expenditures but also rapid methods of replenishing it. It is important to note that bats also know how to save their energy. If desired, they can cool their body almost to the temperature of ice during sleep. Bats always sleep upside down. And they cover themselves with their wings to conserve heat. It is known that bats hibernating in caves during the cold winter months can withstand low temperatures, even when enclosed in ice.

They are immune to bites from even the most venomous scorpion!

Moreover, some bats feed on scorpions. Up to 70% of the diet of pallid bats consists of scorpions! Moreover, scientists have discovered a poison in the saliva of bats, which turned out to be a powerful thrombolytic. With its help, drugs against high blood pressure and stroke can be created, as this poison is absolutely harmless to humans.

They sleep upside down despite gravity

Since bats have disproportionately large wings, it is difficult for them to take off from the ground. That is why they choose resting and sleeping places from where they can start flying simply by loosening the grip of their clawed fingers and spreading their wings. It is amazing that their cardiovascular system has somehow adapted to such an incorrect body position and compensates for the pressure exerted by gravity on the blood flowing to the head during sleep. Bats' heads can rotate 180°.

Huge pulse variability

The pulse of a noctule bat (a genus of bats) is so weak at rest that it reaches 18 beats/min. But during activity, her pulse is 880 beats/min, i.e., it increases almost 50 times! From the perspective of mechanics and material strength, it is unclear how the bat's heart withstands such a pulse. If a human had the same ability to increase their pulse (and accordingly blood flow), then during intense physical exertion, his pulse would reach 3000 beats/min (equivalent to a frequency of 50 Hz).

Professional vocalists

The voice of bats boasts an impressive range of about seven octaves. The sounds produced by bats are inaudible to humans, yet they are incredibly loud. For instance, species such as *Macrophyllum macrophyllum* and *Artibeus jamaicensis* emit sound signals with volumes ranging from 110 to 120 dB. To a human ear, this volume is comparable to the noise of a tractor operating just a meter away. The pain threshold for the human ear is at 130 dB—the roar of an airplane at takeoff. Just imagine the powerful ultrasonic generator housed within such a bat. And it's not just the generator, but also the emitting antenna. Scientists have discovered that bats use their noses to shape a directed beam of ultrasound. Through computer modeling, it was revealed that the tip of the bat's nose is positioned at the focal point of the ultrasonic beam—the point where the signal's propagation lines intersect. Such structure and geometry of the nose serve to maximize the focus of the ultrasonic beam. Moreover, bats are capable of emitting several different signals simultaneously. If we were to compare this to human abilities, imagine a single artist at a concert singing multiple songs at once, each in a different voice. Pure fantasy!

Reader: I think unveiling these secrets of the bat would be a great joy for us, as we could utilize them for the benefit of humankind.

Authors: Undoubtedly. It's no coincidence that in Chinese, the characters for "happiness" and "bat" are written the same.

Chapter 7

We Live on the Charged Sphere

Now entertain conjecture of a time
When creeping murmur and the poring dark
Fills the wide vessel of the universe.

—W. Shakespeare, *Henry V*

We already know that from the moment of the Big Bang, the ship of the Universe sails through the waves of darkness among the voids and black holes, or flies through clusters of bright stars and quasars, all the while being surrounded by magnetic fields. These magnetic fields extend from galactic scales (50 million light-years) to the nano world—the world of membranous ion channels and picoampere currents flowing through them. Interestingly, scientists today are on the path to discovering magnetic fields dating back to the times of the Big Bang. This discovery could change our understanding of the Universe's evolution.

But what about the electric field? Has it remained on the sidelines of life's evolution, while the tree and bushes of evolution flourish in the presence of a magnetic field? No!!!!!!!!!!!

We live in an electric field

The electric field does not miss its chance to affect living beings. Moreover, it can be said that we all live on a charged sphere. Yes,

Shadowless Squids: Stories of Physics in Nature
Vitalii Zablotskii and Tatyana Polyakova
Copyright © 2025 Jenny Stanford Publishing Pte. Ltd.
ISBN 978-981-5129-43-4 (Hardcover), 978-1-003-57062-2 (eBook)
www.jennystanford.com

indeed! Our Earth has a negative electric charge of approximately 600,000 C. This charge is located on its surface and creates a constant electric field with an intensity of about $E = 130$ V/m, which varies depending on the location and weather. The intensity vector of this field is directed vertically downward, so if there is any free electric charge somewhere near the surface, it rushes down, i.e., into the ground. This is utilized in electrical engineering by connecting electrical devices to the ground through a conductor. This is called grounding. Note that when a person walks barefoot, they are also grounded, and the electric charges from them flow into the ground. But we will talk more about this later. For now, let's return to our big charged sphere—Earth.

The "Earth" capacitor

More precisely, we live on one of the plates of a giant spherical capacitor. The capacitor named Earth has two plates: the surface and the ionosphere, which are negatively and positively charged, respectively. As we already said, such a capacitor can accumulate a charge of up to half a million Coulombs.

Who charges the "Earth" capacitor?

The main question is: who or what charges this capacitor? After all, it is charged and periodically partially discharged. Every second, approximately 100 electrical discharges—lightning—occur on Earth! The average current flowing from the ionosphere to the Earth's surface is about 1000 A. If this capacitor were not constantly charged, the Earth's charge would turn to zero in just 500 sec. Indeed, by definition, current $I = q/t$. From here, for $q = 5 \cdot 10^5$ C and $I = 10^3$ A, it's easy to calculate the time $t = 500$ s. But what charges the Earth capacitor?

The global answer is simple: the Sun. Indeed, our Sun is the source of all energy on Earth. The total energy of solar radiation entering the atmosphere in one second is $1.7 \cdot 10^{17}$ W. If we assume that all this power is then emitted by the Earth's surface into the surrounding space (a condition of thermodynamic equilibrium), the average temperature of the Earth's surface would be 293 K, i.e., 20°C. However, as you know, the Earth's surface is uneven, and therefore it heats unevenly. By the way, this is why winds blow. Indeed, wind always blows from an area with higher pressure to an area with lower pressure. And why is the pressure different over different areas of the Earth's surface? It's clear why. Where the Sun heats the surface more, the air temperature rises more, and therefore, the atmospheric pressure increases. And where it's colder, the pressure is lower. It's not for nothing that all hurricanes originate in the tropics, near the equator. There is also a pressure difference between the upper and lower layers of the atmosphere. Therefore, there are both ascending and descending air currents in the atmosphere, which any pilot can tell you about.

Thus, our Sun gives the Earth a tremendous amount of energy, part of which goes to separating charges in the atmosphere.

But how is the spherical capacitor Earth with its plates—the surface and ionosphere—charged? Let's focus on the mechanisms of periodic charging and discharging of this capacitor—the electric cycle in the Earth's atmosphere.

The electric cycle of the Earth's atmosphere

The electric cycle of the Earth's atmosphere consists of several steps. Note that all electrical processes occur in a relatively thin layer of the atmosphere, about 10 km thick.

The first step is the evaporation of water from the surface. Water vapor rises to the upper layers of the atmosphere, where it cools, condenses into droplets and/or turns into ice crystals. Thus, somewhere at high altitudes, an aerosol consisting of water droplets and ice microcrystals forms. Small droplets and ice crystals, being in a suspended state, collide with each other and charge with opposite charges. This is somewhat similar to how an ebonite rod and wool charge through mutual friction. Only in the atmosphere, the role of friction is played by very frequent collisions of aerosol particles with each other.

The second step is the fall of negatively charged microdroplets. At a certain height, these droplets form cumulus clouds. The separation of charges is created by different falling speeds in the cumulus cloud of positively and negatively charged droplets, having different sizes and masses. Large droplets are positively charged, and they fall faster, falling to the ground as rain, or are suspended, forming a local accumulation of positive charges in the cloud. As you see, Earth's gravity also participates in separating charges within the cloud. As a result, a sufficiently large electric charge forms on the lower part of the cumulus cloud. An electric field arises between the Earth's surface and the lower boundary of the cloud. When the intensity of the electric field between the cloud and the surface reaches a critical value, a breakdown occurs—electrical discharge or lightning. The penetration of streams of warm moist air from the Earth's surface into the cumulus cloud facilitates the electrical breakdown of the atmosphere in the form of individual lightning flashes. When electric current flows through the conducting channel of air, the air heats up and expands. Lightning can heat the channel through which it moves to 30,000°C, five times hotter than the surface of the Sun. If the discharge is powerful enough, the speed of air expansion in the channel becomes greater than the speed of sound, creating a shock wave that we hear as thunder. Lightning can occur not only between the cloud and the ground but also between clouds carrying charges of different signs.

Finally, lightning transfers part of the lower (negative) charge of the cloud to the ground. In principle, lightning can also transfer a positive charge to the ground. But, according to measurements, the number of lightning strikes transferring a negative charge to the ground is about twice the number of strikes giving it a positive charge. Thus, it is lightning that transfers the negative charge to the ground and ensures the charging of the Earth capacitor. The average charge transferred by one lightning strike is 20 C. Lightning leads to the growth of droplets and their precipitation on the surface as rain. Thunderstorms are most frequently observed in continental and coastal areas, while far from land and near the Earth's poles, they are virtually absent.

The cloud: A factory for producing electric charges

A thunderstorm cloud is a huge amount of vapor, part of which has condensed into the tiniest droplets or ice flakes. The lower part of the thunderstorm cloud usually carries a negative electric charge, while the upper part carries a positive one. The maximum potential difference, $U = 1.4$ GV $= 1.4 \cdot 10^9$ V, was recorded in thunderstorm clouds at altitudes of 8–10 km in India (GRAPES-3 experiment). The configuration of charges in these clouds was well represented by a flat capacitor with a capacitance of ~0.85 µF (for comparison, planet Earth has an electrical capacity of about 710 µF), and its charging to a voltage of 1.4 GV took 6 min, which (for comparison) corresponds to the power of a large nuclear reactor! And such power is needed to charge just one pair of clouds. Now calculate how many such "nuclear reactors" are needed to keep our planet's charge constant. After all, at any given moment, lightning flashes in more than 2000 storms around the Earth. The current in a lightning discharge on Earth can reach 10–500 thousand amperes. The longest lightning bolt—a mega-lightning bolt 770 km long—was recorded in USA on April 29, 2020. In addition, the longest-lasting lightning bolt in history, lasting 17.1 sec, was registered over Uruguay and northern Argentina.

Interestingly, thunderstorms produce intense gamma-ray flashes lasting up to 1 ms with photon energies up to 100 MeV.

Key words of the electric cycle mechanism in the atmosphere: sun (energy); water (evaporation, condensation, and crystallization); aerosol electrification (collisions and friction); clouds (factories for producing charges); gravity (convection and charge separation), and lightning (charge transfer to the ground).

Lightning and volcanoes

Thus, the separation of charges in the atmosphere and the emergence of the Earth's electric field are determined by the processes of collision and subsequent electrification of aerosols. The same mechanism works in other environments, as electrification can occur during collisions of microparticles of another kind. For example, electrification and lightning can be observed in dust storms, over volcano craters, and during a nuclear explosion. But do not think that scientists fully understand the processes of lightning formation. An interesting discovery was made quite recently during the study of lightning flashes, ash ejections, and lava during the eruption of the Sakurajima volcano in Japan. Mysterious invisible electrical signals occurring in the volcano at the early stages of eruptions were registered. The duration of these high-frequency electrical signals is several seconds, i.e., an eternity compared to the duration of lightning. This discovery may allow scientists to find ways to use high-frequency electrical signals for early warning of impending eruptions.

In conclusion, it can be said that we live on a large, charged sphere. From birth, we find ourselves in an electric field with a strength of $E = (100–300)$ V/m. Moreover, even before birth, from the moment of fertilization of the egg, the electric field left its mark on the development of the embryo. But you will learn about this from subsequent chapters of this book. For now, we will just list some known facts that testify to the role of the electric field in the life of living organisms.

Evolution in an electric field

Do not be surprised, but perhaps lightning is the engine of evolution. In 1953, biochemists Stanley Miller and Harold Urey showed that some of the "building blocks" of life—amino acids—could be obtained by passing an electric discharge through water in which gases of the Earth's "primordial" atmosphere (methane, ammonia, and hydrogen) were dissolved. Fifty years later, other researchers repeated these experiments and obtained the same results. Thus, one of the hypotheses of the origin of life on Earth assigns a key role to lightning.

It is known that 5 out of 21 amino acids have an electric charge. Three of them have a positive charge, and two have a negative charge. Another 6 amino acids are polar, i.e., have an electric dipole moment. Recall that the electric field acts on charges with a force $F = qE$. Moreover, the force acting on a positive charge is directed along the field, and on a negative charge—opposite to the direction of the field. The electric field acts on an electric dipole similarly to how a magnetic field acts on a magnetic dipole (magnetic needle): the dipole is oriented along the direction of the field. Thus, in a living organism's cell, 11 out of 21 amino acids are constantly under the influence of the Earth's electric field: they are constantly subjected to electric forces or rotating moments. Can we think that the evolution of life on Earth overlooked the action of these forces on the molecular "building blocks" of life—amino acids? Apparently not!

Another example. When short pulses of current are passed through a bacterium, pores form in its membrane, through which fragments of DNA from other bacteria can pass, triggering one of the mechanisms of evolution.

Probably, you will not dispute the fact that primitive people walked barefoot. But now you know that they walked on negative electric charges. Considering that wet human skin is a conductor of electric current, it can be said that the negative charge of the Earth's surface was transferred to the person, which is equivalent to the flow of positive charge from the person to the ground. Perhaps evolution somehow took this fact into account and made appropriate adjustments in the human body. But modern people wear shoes with dielectric soles, thus isolating themselves from the charges of the Earth. It's hard to say whether this is good or bad. But it is well known that the well-being of a not entirely healthy person depends on the local intensity of the Earth's field, as well as on changes in atmospheric pressure, in most cases accompanying changes in field intensity.

Here we will not tire the reader with a full review of the possible effects of the electric field on the evolution of living organisms. Probably, our reader is already tired and wants to take a shower. And as is well known, a shower relieves fatigue. And do you know why? During the day, during physical activity, positive charges, mainly calcium ions involved in muscle contraction, accumulate in the body. In the shower, since water is a conductor of current, though a poor one, these charges flow into the ground. Perhaps it is precisely because of such grounding that a person begins to feel more energetic. And if you have listened to our advice and already received a charge of vigor, shedding the excess electrical charges from yourself, you can proceed to read the next story.

Chapter 8

Where Hides the Living Electricity

Smith, a journalist: Mr. Tesla, aren't you a bit biased
towards electricity?
Nikola Tesla: Electricity is me. Or, if you prefer,
I am electricity in human form. Mr. Smith, you too are electricity;
you just don't realize it.

Nikola Tesla was right when he said that he himself is electricity. We will begin our story about bioelectricity from the top down, i.e., from large scales, gradually descending to molecular sizes. You already know what an electric charge is and its properties. As we have described in the previous story, we all live inside a large spherical capacitor named Earth, which is constantly being charged and periodically discharged through lightning. No one can deny that electrical charges and atmospheric currents affect living organisms. On the organism level, the approach of a storm is felt for many hundreds of kilometers, giving animals time to find shelter from lightning and rain. By the way, if you ever find yourself in a place where your hair suddenly stands on end, leave that place immediately. This happens before a storm and means that this part of the Earth's surface has a sufficiently large electric charge, indicating that lightning is very likely to strike exactly in that place. Many people sensitive to weather can feel hurricanes (which usually

Shadowless Squids: Stories of Physics in Nature
Vitalii Zablotskii and Tatyana Polyakova
Copyright © 2025 Jenny Stanford Publishing Pte. Ltd.
ISBN 978-981-5129-43-4 (Hardcover), 978-1-003-57062-2 (eBook)
www.jennystanford.com

carry a huge electric charge) from many hundreds of kilometers away. But we won't delve into this topic here; instead, we'll look at the electrical currents and fields of living organisms. Let's ask a direct question. Are there electric currents inside living organisms?

Yes, of course. In the body, electrical currents flow from the formation of the embryo, immediately beginning to play a significant role in its development. For example, during the embryonic development of the embryo, the asymmetry of the left and right is created by an electric current running along the length of the notochord, generating a magnetic field vector directed either to the right or to the left. In adults, pulses of electric current control the workings of the heart. Here's an example from everyday life. When we touch something very hot, like a candle flame, we instantly pull our hand back. How can this be explained in terms of electricity? It's both simple and complex at the same time. So, you accidentally touched the flame. What happened next? There are so-called thermosensitive ion channels on the skin that open in response to a rise in temperature, thereby changing the cellular membrane potential and generating an electric pulse in the form of a potential spike. This pulse spreads along the nerve fibers like through wires to a special area of the spinal cord. Here, the received electric signal is processed and identified as a burn hazard. As a result, the spinal cord generates a return electric pulse, which in the internal language means: "urgently pull the hand back." This signal is sent as quickly as possible through other nerves going from the spinal cord to the arm muscles. Upon receiving this electric impulse, the muscle contracts sharply, and the hand jerks away from the flame. For more clarity, let's break down the step-by-step mechanism of pulling the hand away from the flame.

It looks like this: dangerous increase in skin temperature >>> opening of ion channels (leading to very small currents, on the order of a few *pA*, passing through the cell membrane) >>> change in the cell membrane's electric potential and formation of an action potential >>> spread of the electric impulse through nerve cells to the spinal cord >>> signal processing, generation, and sending of a return electric impulse strictly to the address of the required muscle >>> spread of the electric impulse through nerves to this muscle >>> muscle contraction. And that's not all. Because the muscle contraction, in turn, means that positive Ca^{2+} ions come out of their depots and literally leap between negatively charged lobes inside the muscle. And as a result of the Coulomb attraction between these negative and positive charges, the muscle contracts.

What do we have in the end? A simple movement—pulling the hand away from the flame—triggered an entire cascade of electrical processes on scales from nanometers (the size of an ion channel) to meters (the total length of the nerves—conductors of electric current involved in the process of transmitting the impulse back and forth). Meanwhile, the speed of the spreading nerve impulse can reach 100 m/s (=360 km/h). Such a high speed of transmitting the electric impulse is necessary for us to stay alive in an emergency. Here, of course, we rely not only on the high conductivity of nerve tissue but also on the quick processing of information. Interestingly, much of our nervous system works without our knowledge, in automatic mode, controlling not only extreme situations but also our heartbeat, blood pressure, breathing, and other body systems.

Let's touch upon more complex processes, not automatic ones, e.g., thinking. It is considered that humans think using the brain. In the human brain, there are approximately 100 billion neurons, which form an incredibly complex electrical network. When the brain is active (e.g., a person is thinking about where and how to move an object), a large part of these neurons are electrically excited, and in search of an answer, they exchange electric impulses with each other. When a decision is made, it is sent as an electric impulse to a specific brain center (for approval), and from there, a muscle receives an electric order for execution. But this is just in broad terms. What really happens in the brain, its spatial and temporal hierarchy of sparks and even micro-lightnings flashing between neurons, we can only guess. And this is despite the fact that the electrical activity of neurons can be visualized using computer tomography or magnetic resonance imaging. Find these images on the internet and see for yourself that this represents the highest level of organization of living matter and its symbiosis with electricity. Much here is unclear. But one thing is clear: no electricity—no life.

But where does the living electricity hide?

It hides in the cell. More precisely, in the cell membrane. The cell membrane is a phospholipid layer, the inside of which is negatively charged, and the outside is positively charged. In other words, the cell membrane is a microcapacitor. Remember, we live in a capacitor named Earth, and now we know that each of our cells has its miniature capacitor. But what charges or discharges the cellular capacitor? Are there also lightning strikes there? No, in the membrane, there are ion channels and ion pumps that maintain the difference in electric potentials between its inner and outer surfaces, constantly redistributing positive and negative ions across both membranes, allowing ions to enter and exit the cell. The resulting potential difference or electric voltage in biology is called the membrane potential. Note that overall, the cell is electrically neutral, i.e., the sum of all its negative charges equals the sum of the positive ones. For different types of cells, the membrane potential varies from −5 mV (in embryonic, stem, and cancer cells) to −100 mV in neurons and cells of skeletal muscle tissue.

The ability of cells to maintain and regulate their membrane potential is crucial for many processes, including the regulation of cell volume, the cell cycle, DNA synthesis, stem cell differentiation, proliferation, muscle contraction, signal transmission, the spread of cancer cells, and wound healing. While it is well-known that membrane potential and bioelectric signals significantly control cell behavior, many mysteries remain. For example, it's unclear how undifferentiated and cancer cells manage to maintain a very low membrane potential, allowing them to be mitotically active (i.e., divide) and highly plastic. In contrast, mature, fully differentiated cells tend to be hyperpolarized (have a large membrane potential) and usually do not undergo mitosis (Fig. 8.1).

As mentioned earlier, the cooperative action of ion channels in the cell membrane establishes the electrical charge of the membrane and maintains the membrane potential. But if the membrane potential is lost for some time, cells always commit suicide (apoptosis). The most astonishing part is the structure of ion channels. Imagine a lipid membrane 4–5 nm thick, embedded with special proteins with pores approximately the size of one of the ions (0.2–0.7 nm): K^+, Na^+, Cl^-, or Ca^{2+}. These channels are not just embedded in the membrane; they can move across it and gather where needed. A single cell's membrane may contain up to 10,000 ion channels. Moreover, channels can open or close upon specific commands, which can be mechanical force or stress, a local change in electrical

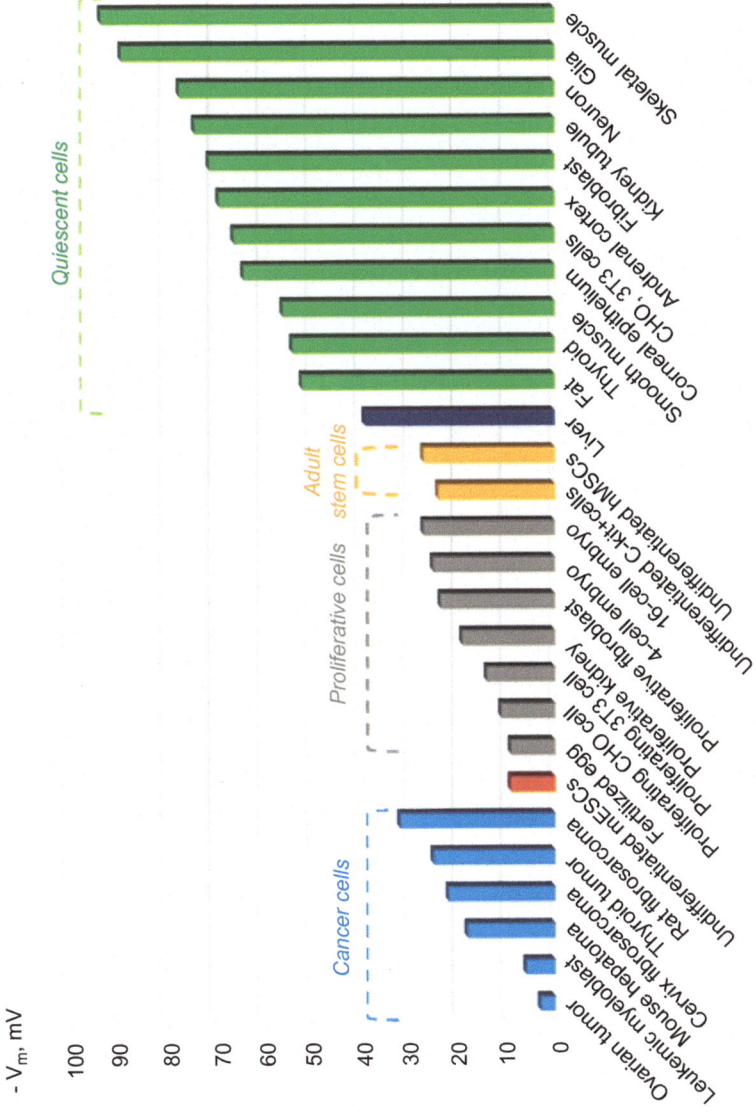

Figure 8.1 Membrane potentials (in mV) for different types of cells.

potential, a change in temperature, or binding of the channel with certain types of molecules (binding with a chemical messenger). Essentially, all controlled channels convert thermal, chemical, or mechanical signals received by the cell into electrical ones. There are also uncontrolled channels and dual-controlled channels, which are opened or controlled both by ligands and electrical potential. All these channels work together to maintain the electrical voltage between the inner and outer surfaces of the cell membrane. Since electrical currents flow through the active channels, ion channels create small magnetic fields around themselves, about 1 nT. Thus, even at the subcellular level, life exists alongside not just electrical but also magnetic fields. Interestingly, the electric field strength in the cell membrane is quite high, $E \approx (10^6 - 10^7)$ V/m.

So, the cooperative action of ion channels regulates the charge of our cellular capacitors—the membranes. The life and fate of a cell largely depend on the magnitude of its membrane potential or the electrical charge of its membrane.

Now, think about something. Or, as they say, rack your brains. Did you? Can you feel how millions of ion channels opened in your head, significantly increasing the electrical activity of neurons?

Ralph: Are you trying to say that thought is material?

Authors: Ask our household philosopher about that.

Diogenes: What can I say here? Of course, in time and space, thought correlates with the electrical activity of brain neurons. But to conclude that thought is just electrical charges zipping around in our heads, I wouldn't go that far. Perhaps there's something else we don't understand.

Authors: We might agree with our esteemed philosopher's opinion.

Ralph: I have another question. So, a living cell is the smallest accumulator and source of electricity in the body. Is that correct, or is there something even smaller where live electricity hides?

Authors: You're right. There's something even smaller— mitochondria. They live inside the cell and have their own membranes. The mitochondrion's membrane is also permeated with ion channels and has a membrane potential.

Chapter 9

A Magical Journey Inside a Living Cell

A kind wizard offered to shrink me down to the size of a protein (say, 50 nm) and send me on a journey inside a living cell. Remembering the folk wisdom that "seeing is believing," I immediately agreed.

Oh, wow... Where have I ended up? It seems I'm floating in some kind of viscous liquid, observing strange creatures and unusual, working mechanisms. Oh, how I regret not bringing a camera or a mobile phone with me! Therefore, I'll describe in words what I see around me.

I'm slowly floating in a slightly viscous liquid that smells pleasant (this is the cytosol) among mysterious mechanisms and a web of microtubules, feeling almost weightless, like an astronaut on a space station. A soft, dim yellow light surrounds me. Only occasionally, against the yellow horizon, can bright flashes of blue light be seen. What could they be? Biophotons? Breathing is very easy, and there's a light scent of blossoming trees. Could this be what the inside of a cell looks like? It all resembles a magical fairy tale with kind dragons and exotic mechanisms producing a barely audible, monotonous hum.

Suddenly, I saw a beautiful girl floating toward me. She must be a princess, I thought, because every fairy tale should have one.

Shadowless Squids: Stories of Physics in Nature
Vitalii Zablotskii and Tatyana Polyakova
Copyright © 2025 Jenny Stanford Publishing Pte. Ltd.
ISBN 978-981-5129-43-4 (Hardcover), 978-1-003-57062-2 (eBook)
www.jennystanford.com

"Hello," she said.

"My name is XX. I will be your guide. You can ask me questions."

"I've always dreamed of a guide-princess," I exclaimed.

In a moment, I felt caught in a mild current and decided to go with the flow for a while. Nearby, small bubbles were transporting nutrients necessary for the life of the cell. The vesicle is separated from the cytosol by a minimal lipid layer, so I could see proteins and other molecules inside them. It looked as if small, transparent, spherical trucks were bustling to and from inside the cell.

After some effort with my hands and legs, like an experienced pearl diver, I left the current and headed toward a sufficiently large object that from afar resembled a large factory. As I got closer, I recognized it as a mitochondrion—a double-membraned spherical or ellipsoidal organelle about 1 μm in diameter. We were taught about them in biology classes at school. But what I saw in reality was shocking. It's the power station of the cell. More precisely, one of its power stations. Just like a conventional thermal power station built by human hands, it serves to generate heat and electricity.

In thermal power plants, the source of energy is the combustion of gas, coal, or oil products. Combustion is defined as a reaction between an oxidizer (often oxygen) and a fuel. Similarly, the mitochondrion also utilizes oxidative chemical reactions—reactions where a substance combines with oxygen, during which the atoms of one substance (the oxidizer) donate their electrons to the atoms of another substance. In the mitochondrion, the oxidation of organic compounds occurs, and the energy released is used to generate electrical voltage (potential), synthesize ATP, and maintain the normal temperature of the cell. Oh, I see how, beneath the inner membrane, along a chain made of proteins, electrons jump. It's simply fantastical to witness what an ordinary person cannot see but is accessible to a tiny human 50 nm tall. And don't ask me what color an electron is. It's like asking what color the clouds in the sky are. I see the jumps of tiny electron clouds inside the mitochondrion.

"Why are mitochondria surrounded by a double membrane?" I asked.

"Well, there's a high electrical voltage there, up to a hundred millivolts. It must be enclosed by a fence for the safety of its surroundings," XX replied.

"Oh, what's that? It's like some kind of mushroom sprouting through the inner membrane of the mitochondrion. No, it's more like a working electric motor!" I exclaimed.

"This amazing contraption is called the proton ATP synthase (ATP-synthase protein complex)—the smallest electric motor in nature, only 10 nm in size. With its help, cells produce adenosine triphosphate (ATP)—the substance that serves as the main source of energy in the cell (ATP mass density 1.04 g/cm^3 and molecular weight 507.18 g/mol). And it works somewhat like this. Protons accumulate in the intermembrane space and then rush back into the mitochondrion, creating an electric current. Protons pass through special channels in the ATP synthase, which is embedded in the inner membrane. This flow of protons spins a special rotor in the ATP synthase, like a stream of water turning a watermill. The rotor rotates at a speed of 300 revolutions/sec, comparable to the maximum revolutions of a Formula 1 car engine. And it's this rotation that leads to the formation of the high-energy molecule— ATP. As it rotates, each of the three catalytic centers of the rotor becomes active in turn. And at the moment of its activity, it attracts an ATP molecule and a molecule of phosphoric acid, which, being so close to each other, react to form an ATP molecule and a molecule of water. It's estimated that an adult human synthesizes and consumes about 40–50 kg of ATP per day, with each molecule's life being very short. Considering that a human is made up of 10^{14} cells, on average, one cell produces about $5 \cdot 10^{-10}$ grams or $3 \cdot 10^{12}$ molecules of ATP per day. This frenzied synthesis of ATP is necessary for energy production and work performance. In other words, mitochondria constantly produce and stockpile ATP "firewood," which the cell transports and burns in the necessary places, obtaining heat and energy to perform work.

My hair stood on end from the static electricity near the mitochondrion. It was time to get out of there. I grabbed onto a thin tube and began to ascend. But the tube disintegrated right before my eyes. I barely managed to grab onto another. Ah, this must be the cytoskeleton, I thought, recalling a biology lesson from school. These filaments and tubes, thinner than my body, about 25 nm in thickness, support the entire body of the cell. But the strangest thing is that these microtubules can very quickly disassemble into monomers

(5 nm long) at one place and then just as quickly reassemble in another part of the cell. In biology, these processes are called the depolymerization and polymerization of F-actin. For instance, the speed of assembly/disassembly of an actin filament is on the order of hundreds of monomers per second, roughly corresponding to 500 nm/s or 30 μm/min.

Interestingly, when a cell wants to change the direction of its movement or its shape, it does not move the cytoskeleton tubes but simply disassembles them in one place and quickly reassembles them in another needed location. Have you seen anything like this in our lives or technology created by humans? It's simply impossible to design a mechanism or moving machine in which the main frame (such as the chassis in a car) is constantly disassembled and reassembled elsewhere. And all of this is done automatically, under the control of only the laws of chemistry and physics.

"Oh..., oh, who is that walking along the actin filament? Walking leisurely on two legs as if climbing a ladder, just like a human, and it seems to be carrying something."

"That is one of the molecular motors—myosin V. The walking protein, myosin V, can actively move along actin fibers and drag attached loads. Each step of myosin V begins with one of its 'legs' (the rear one) detaching from the actin filament. Then the second leg leans forward, and the first freely rotates on the 'hinge' connecting the legs of the molecule, until it accidentally touches the actin filament and sticks to it."

"And by what forces does the myosin leg attach to the actin filament? Does it have claws on its legs like a squirrel, to hold onto any surface irregularities?"

"No, of course, not claws. The attraction forces between the electrical charges on the myosin leg and the actin filament allow the actin legs to attach to the filament. But let's move on. We shouldn't hinder the myosins from delivering their loads," said XX.

"And where do they transport the loads?"

"From the cell membrane to the cell organelles, they deliver nutrients and other necessary elements, and waste and various unnecessary substances are transported back to the membrane."

"You reminded me of nutrients. And I've already become quite hungry. After all, I've expended a lot of energy, making my way through the cytoskeleton and a forest of various mechanisms. Is there a cafeteria here or something similar?" I asked.

"There's no cafeteria here. But here, take this and eat a little," she said, grabbing a couple of glucose molecules floating by us.

"Can this really be eaten?" I asked.

"Glucose is the main food product and energy source for the cell. Take this, it's D-glucose, $C_6H_{12}O_6$—one of the most common energy sources in living organisms on the planet. Eat!"

"What if I want L-glucose," I said sarcastically.

"Oh, you know that glucose molecules are chiral," XX was surprised. "Yes, there is its left enantiomer (L-glucose), which is practically not found in nature. So, eat what's available."

I tried a molecule of glucose and found it to be sweet.

"Maybe I'll save these molecules for dessert, and now I'll eat some protein idly floating in the cytosol?" I asked.

"There are no proteins idly floating in the cell. All of them move in strictly defined directions. And how will you eat it? Look, the proteins are about the same size as you," said XX.

"Well, okay," I said and started nibbling on a glucose molecule.

Replenished, we moved on.

And then we approached something large and solid.

"This is the nucleus of the cell," XX told me. "But we won't go in there," she added.

"Why not?" I asked.

XX paused and answered: "It stores DNA, and its replication occurs there. Life is born there, and thus it's a sacred place for all the inhabitants of the cell. We cannot interfere with DNA synthesis. If an error occurs during its synthesis, e.g., due to a virus that has illegally penetrated the nucleus, it could affect not only the future of the living organism but also its subsequent generations. The nucleus carefully guards its chromosomes, and therefore its mechanical strength is quite significant. Chromosomes are the only objects in the cell capable of replicating. And the life of every new creature starts with them."

"Can I at least peek into the nucleus through this small hole or window? I really want to see DNA replication."

"Yes, but only for a short while."

I cautiously peered through the window and instantly recoiled in horror.

"What did you see?" asked a surprised XX.

"I saw two intertwined vipers performing a mating dance. Once, I saw this during a trip to Armenia. Two enormous, two-meter-long vipers, intertwining, untwining, and intertwining again, performed a mating ritual. It was a terrifying and unforgettable sight!"

"There are no vipers here, although DNA, like a viper, is 2 m long when fully unwound. You're seeing DNA synthesis. Keep watching," XX reassured me.

I looked through the window again timidly. There, a double helix, spinning rapidly, separated into two single strands. And immediately, on each resulting strand, a second strand was built, forming two identical daughter DNA molecules, which then twisted into separate spirals. This was done by a complex consisting of 15–20 different protein enzymes, called a replisome.

"Are you surprised?" XX asked.

"Yes. The synthesis process is extremely fast, DNA rotates at a furious angular velocity, hundreds and even thousands of revolutions per minute. It's as fast as the shaft in a car engine rotates."

"And what did you think? A single DNA contains a vast amount of information. That's why the process of transcribing information during DNA synthesis must be very fast. Yet, even at this speed, the information copying process is relatively slow. For example, the bacterium E-coli (genome length $5 \cdot 10^6$ bp, replication speed <1000 bp/s per replisome) copies and replicates its genome in about 40 min."

"All this is extremely interesting," I said. "But let's swim and see what else is there."

Next to the nucleus, I saw a complex branching system of cavities, vesicles, and tubules, surrounded by membranes. It looked like a sophisticated system of screens hiding something important from prying eyes.

"This is the endoplasmic reticulum (ER)—one of the cell's organelles," XX said.

The cavities of the endoplasmic reticulum (ER) open into the intermembrane space of the nuclear envelope. Different types of endoplasmic reticulum perform various functions in the cell, from the active transport of many elements into the nucleus to the synthesis of hormones and the accumulation and transformation of carbohydrates.

The endoplasmic reticulum also serves as a depot for calcium, which is, among other things, a mediator of muscle cell contraction. The concentration of calcium ions in the ER can reach 10^{-3} mol, while in the cytosol, it is around 10^{-7} mol (in a resting state). Under the action of some other stimuli, calcium is released from the ER through facilitated diffusion. The return of calcium to the ER is ensured by active transport.

"And why does the cell need to accumulate and store so many calcium ions? After all, the charge of each calcium ion is equal to the charge of two electrons. And if you gather many such ions in one place, it will create a cloud of large positive charge. It's like a dark thundercloud in the atmosphere. Does the cell need this?" I asked.

XX thought for a moment and said: "You're right, the calcium ion depot is a large electrical charge. I don't know how the cell holds it, as like charges, as known, repel each other. Apparently, this is why the ER has a system of vesicles and tubules, separated by special membrane partitions. This calcium ion reserve is vital for the cell. The concentration of calcium ions in the cytosol affects many intracellular and intercellular processes, such as the activation or inactivation of enzymes, gene expression, synaptic plasticity of neurons, muscle cell contractions, and the release of antibodies from immune system cells."

Right next to the ER, we saw the Golgi apparatus. The body of this organelle looks like a stack of membranes or flat sacs placed on top of each other. It's like a packaging factory that aids in the secretion of substances synthesized in the endoplasmic reticulum, processing and packaging of proteins and lipid molecules, as well as proteins intended for export from the cell. Almost all substances secreted by the cell (both proteinaceous and non-proteinaceous) pass through the Golgi apparatus and are packaged there into secretory vesicles.

"Is this a transit point for molecules and proteins?"

"Something like that. Here, substances accumulate and are sorted, modified, and packaged before being sent to their destinations," XX replied.

"There must be very complex logistics there," I noted.

"Undoubtedly, otherwise it all wouldn't be able to function correctly," XX replied.

"Oh, oh! I think it mistook me for some protein and wants to package me and send me somewhere!"

"Yes, we need to get out of here quickly," XX responded.

We moved on. Before us was a lysosome—a membrane-bound cellular organelle, usually less than 1 μm in size, inside which an acidic environment is maintained, and contains many hydrolytic enzymes dissolved in it. The lysosome is responsible for the intracellular digestion of macromolecules, including autophagy (a natural, regulated mechanism of the cell that disposes of unnecessary or dysfunctional components). Normally, the pH in lysosomes is about 4.5–5, meaning the concentration of hydrogen ions in them is two orders of magnitude higher than in the cytoplasm. This is maintained by the active transport of protons, performed by a protein pump embedded in the lysosome membranes—vacuolar ATPases. In addition to the proton pump, the lysosome membrane incorporates transporter proteins for the cytoplasmic transport of hydrolysis products of macromolecules: amino acids, sugars, nucleotides, and lipids.

"If I understand correctly, the cell digests (breaks down) everything it doesn't need in the lysosomes?" I asked.

"Yes, that's one of the functions of a lysosome. For example, if something foreign and inedible enters the cell, it sends it to one of the lysosomes for digestion," XX answered.

"Are you hinting at me? After all, I illegally penetrated the cell and am obviously not needed by it."

"Yes, that's what would have happened to you. But don't worry, you're with me," XX reassured me.

"Thank you, my guardian angel. But let's move away from these lysosomes."

"And now we'll look at the cell membrane," said XX.

Navigating through the labyrinths of microtubules, we reached the cell membrane. It seemed to me that the membrane consisted of two layers of little heads arranged tail-to-tail, with their heads forming the inner and outer surfaces of the membrane. The membrane was only 4–5 nm thick. The membrane seemed to breathe, bending slightly now in one direction, now in another. In addition, light waves ran across it, rocking the proteins embedded in the membrane.

The lipid membrane of the cell is a very important element. The cell has no eyes, so it senses and perceives the external world through its membrane. It's probably similar to how we feel touches or warmth (cold) with our skin. The membrane has thousands of different types of receptors and channels. Ionic channels are special proteins with very precisely calibrated openings, floating in the lipid membrane. The diameter of the channel is so precisely calibrated that it can allow only one type of ion to pass, i.e., its diameter matches the size of the given ion. Different ionic channels allow the passage of Na^+ (sodium), K^+ (potassium), Cl^- (chloride), and Ca^{2+} (calcium) ions. Ionic channels can be uncontrolled or controlled. The gates of a controlled channel can close and open on a specific command.

"Who gives these commands?" I asked.

"The controlling commands can come from the cell itself as well as from other cells," XX replied.

The most common types of channels are ligand-gated ion channels and voltage-gated ion channels, which open (or close) with changes in the electrical voltage across the membrane (membrane potential). Ligand-gated channels convert chemical signals received by the cell into electrical ones; they are necessary, in particular, for the functioning of chemical synapses. Voltage-gated channels are needed for the propagation of action potentials.

There are also thermosensitive channels, which open or close with just a fraction of a degree change in the membrane's temperature. Many ion channels are very sensitive to the slightest mechanical stresses ($\sim$0.1 Pa).

Thanks to the controlled passage of sodium, potassium, chloride, and calcium ions through their respective channels, the cell can change its membrane potential (the electrical voltage between the outer and inner surfaces of the membrane) from -70 mV to $+100$ mV. The intensity of the electric field in the membrane can reach 10^7 V/m. For comparison, the intensity of the electric field between the two contacts of a household socket (220 V) is about 10^4 V/m, i.e., a thousand times less.

"So, this is where electricity hides!" I exclaimed.

"Yes, here in the cell, many things work on electricity," XX replied.

"I see how electrical currents flow through the ion channels. But how large are these currents?"

"Ionic currents are small: from 1 pA to a hundred pA (picoamperes). But consider that the outer membrane of a single cell can have up to 10,000 different channels. And there are also ion channels on the membranes of mitochondria. And all this works simultaneously. So, the cell is permeated with electrical currents. Nikola Tesla was right when he said, 'Electricity is me.'"

"But I learned that an electric current creates a magnetic field around itself. Does that mean there should be a magnetic field near the channels?"

"Correct, the magnetic field of working channels is registered by special ultra-precise magnetometers. This field is small, on the order of tens or hundreds of pico-Teslas (pT). But it exists and can be measured!" XX replied.

The lipid membrane is probably the softest material among all known natural materials. For example, the membrane of nerve cells has an extremely low Young's modulus ($E = 100$ Pa), the smallest of all known materials. This extraordinary softness of the cell membrane allows it to serve as the sensory organ of the cell. After all, the cell primarily senses the external world through mechanical forces or stresses applied to its membrane. Thanks to the remarkable softness of the membrane, ion channels and receptors can move along its surface, forming sufficiently large clusters in those parts of the membrane where the cell needs them at the moment.

"Can I get closer and touch the membrane? I want to try what it feels like and take a closer look at the channels."

"Yes, you can carefully approach it. But be careful not to trip over the actin cortex, which is a specialized layer of cytoplasmic proteins on the inside of the cell membrane," XX responded.

I approached the inner part of the cell membrane and touched it with my hand. Then something unexpected happened! A small part of the membrane, slightly larger than my height, bent outward into the extracellular space, and I found myself inside a hemisphere made of the lipid envelope. Then, the hemisphere detached its ends from the rest of the membrane and quickly closed, turning into a complete spherical shell. In an instant, I was in this transparent sphere outside the cell.

"This is exocytosis—the process of moving substances from the cell into the external environment," XX yelled to me and waved goodbye. I also waved back and thought that my journey had safely come to an end. But I was mistaken.

It turned out that in this transparent sphere, I had entered the intercellular space. The first thing I saw was a huge cell with appendages in front. It worked with these appendages like a snowplow, scooping up everything in its path. This was a macrophage—an immune system cell. The name macrophage speaks for itself: macro—large, phage—from the Ancient Greek φἄγω "I eat," i.e., large eater. Macrophages devour harmful bacteria and viruses, thereby ensuring the organism's victory over infection. It seemed to me that the macrophage was chasing me, mistaking me for some pathogenic organism. I was trembling with fear, anticipating that it would soon catch my transparent sphere and grab me with its terrifying appendages. But fortunately, it turned out that the macrophage was not after me, but after an E-coli bacterium. It spun its flagellum like a propeller at a frantic speed, constantly changing directions to save its life. But the macrophage was quicker. Before my eyes, it captured it with its huge appendages and sent it directly inside itself. It ate her! But it had neither a mouth nor teeth. Simply, an opening appeared in its body, swallowed the bacterium, and closed. What happened next is simply too horrifying to tell. You wouldn't see it even in a horror film. Through the transparent wall of the macrophage, I saw the outlines of the bacterium blur. It dissolved her in acid!

I had to run away from this terrible acid tank quickly! And then, to my luck, I caught a current and was safely carried away from this monster. "This is an intercellular river," I thought and regretted not having a guide assistant like XX nearby. Small molecules and various proteins floated by me. These are signaling molecules—messages in bottles that cells send to each other.

But what's this? Ah, the fire-breathing dragon that must be in every fairy tale. In my path was a powerful stream of charged particles, under whose influence my hair stood on end either from fear or from electricity. Approaching, I saw that it was a stream of calcium ions (Ca^{2+}), which the cell expelled from itself like a geyser ejects hot water and sent it as an electrical signal to other cells. Such an electrical signal is faster than signaling molecules spreading by diffusion, at a speed, roughly, of letters in bottles carried by the will of the waves. And calcium signals are more advanced; they spread through neurons as pulses of electric current, I thought.

Safely passing the powerful stream of calcium ions, I found myself in a relatively calm place and decided that all dangers were behind me. But suddenly, my attention was drawn to a group of aggressively disposed cells called T-killers (from English killer "killer") or cytotoxic T-lymphocytes, whose main function is to destroy damaged cells of their own organism.

That's exactly what they were doing, surrounding an unfortunate cell. And around, like hungry hyenas, stood phagocytes—immune system cells that protect the organism by engulfing (phagocytosis) harmful foreign particles (bacteria, viruses), as well as dead or dying cells. This time, the victim of the T-killers was a viral factory—a cell infected by a virus, which still continued to produce and release thousands of virus copies into the intercellular space per second.

How did these T-killers find the infected cell? Quite simply. The sick cell, not wanting to infect its friends and neighbors, sends molecular signals like "I am captured by a virus, kill me." This can be called an anti-SOS signal (for those who don't know: SOS—"Save Our Souls," used in maritime since 1905 by sailors experiencing a shipwreck as a signal-request for urgent help). But the infected cell is not asking for help for itself. Viruses, having penetrated the cell, multiply in it until the cell dies. That's why the cell sends such suicidal, but salvational for other cells signals. Receiving the signal "...kill me" and recognizing it with special receptors, the killers find the cell that sent this signal and kill it. Such a killer is a serial killer, it can strike 30–40 cells in a row.

But how do T-killers kill the cell? Do they have special weapons? Yes, they do. It all happens like in a classic detective story. The T-killer carries special granules, in which deadly toxins are hidden (the killer cell itself is protected from these toxins). Using special molecular sensors, the killer cell recognizes infected or cancerous cells to be eliminated. Finally, the killer cell presses against the victim cell, leaving only a small gap between their membranes, into which it releases poison, killing the unfortunate exhausted cell. A typical killer has three poisons or three proteins: perforin, granzyme, and granulysin. Perforin acts first. It integrates into the outer membrane of the victim cell and forms pores in the membrane, through which granzyme and granulysin pass. Entering the cell, these toxins trigger the mechanism of apoptosis—cell self-destruction. Within 10–20 min, the cell digests its content into the simplest molecules and packages these safe remains into small bags, which phagocytes eagerly eat. That's how T-killers work: very neatly, without harming neighboring healthy cells.

Figure 9.1 Do you see the most important molecule here?

Interestingly, newly born T-cells are not able and not ready to kill. And few of them are born. They grow like ordinary peace-loving cells, but when their sensors encounter an infected or cancerous cell, T-cells begin to proliferate intensively and simultaneously acquire the properties of professional T-cell killers. This process of proliferation and transformation takes days, if not weeks.

Well, while I've been telling you all this, the infected cell was killed, its remains were eaten by phagocytes, and the T-killers went in search of a new victim.

Good that they didn't notice me. But I really didn't send any anti-SOS signals, so why would they kill me. Although phagocytes could easily eat me alive, as in their understanding I am a foreign body. But I fled in time.

It feels like it's time for me to leave this intracellular and intercellular world. It's time for me to transform back into a normal human. I found a piece of paper in my pocket with the most complex ancient spells that the wizard had provided me, and I started to mutter them. And just in time, a snowplow machine named macrophage was already approaching me.

The spells transformed me back into a normal human, and I began to summarize this unusual journey.

Journey summary

First, what shocked me was the intracellular electricity: Ca ion storages, charged membranes, electrical voltage, currents in ion channels, electromotors powered by pure protons, myosins with charged legs walking on actins, electrical signals between cells, and just ions dissolved in the cytosol. And add to this, electrically excitable cells, like nerve and muscle cells. The latter, under the influence of an electrical impulse (action potential), change their length at a frequency of about 1 Hz. In the end, I would say that a cell is primarily an electrical system. Electrical charges, currents, and forces play one of the main roles in the cellular machine. One could say a living cell is an electrical unit!

Second (or maybe first) shocking and mysterious thing is the logistics of the entire cellular machine. What (or Who) knows where and when to transport specific molecules and proteins? Where are these route lists written? There's no answer, but I saw how all this works flawlessly as long as the cell and organism are healthy. Of course, it can be said that all processes in the cell are governed by the laws of physics, chemistry, and biology. And this would be true.

But, of course, not the whole truth. Nothing in our Universe escapes the action of these laws. But life, at least at the cellular level, is not everywhere.

Third, is the extraordinary complexity and at the same time the precision of the work of individual cell mechanisms, as well as the whole cell. Have you seen how an internal combustion engine of a car is designed, in which many parts and surfaces are calibrated, adjusted, and processed with precision to the micrometer? In the cell, all parts are calibrated with precision to the size of a single atom! And how many years can a car engine last? Say 10 or even 20 years. And how many years do cells live and work? If placed in favorable conditions and fed, individual cells can live forever. This is widely used in biotechnologies. Many cell lines start their history from the middle of the last century when they were first isolated, e.g., from human embryos.

Reader: Stop, stop. It seems to me that by comparing the service life of a car engine with the lifespan of a cell line, you are being unfair. After all, neither a car nor its engine has the ability to self-replicate through division.

Authors: You are absolutely right. With this example, we wanted to emphasize the distinctive feature of the complexity of the cellular machine—the ability to self-reproduce. After all, it wouldn't be bad for a car to learn this. One might say that we inadvertently looked into the future of car manufacturing ☺. But let's return to the facts that shocked our cell traveler.

Killings and intercellular wars! It turned out that cells also wage war against each other. And this war is constant, with no beginning and no end. Some, through mutations, perfect their weapons of attack, while others constantly search for new ways of defense. The main weapons of the defending cells are poisons and acid. But mechanical means of fighting against enemy forces are also known. It was recently discovered that mechanical pressure exerted by healthy cells on an area occupied by tumor cells can stop the proliferation of cancer cells and prevent their growth. Perhaps cells have some secret weapon, but I couldn't see it. Especially since I wasn't allowed everywhere. But I'm thankful for what I did see.

And I also saw that cells communicate with each other. Cells talk to each other in 5 different languages. What languages do cells use for communication? Here they are five: chemical (signaling molecules), tactile (touching each other and transferring signals through forces applied to the membrane), electrical (using calcium ions), light (these are biophotons), and finally, sound (yes, yes, some cells can emit acoustic waves). And cell receptors are translators, translating from these 5 languages into cellular.

In conclusion, I can say that this journey gave me only some understanding of what happens in a living cell, but even more remains unclear. But as George Bernard Shaw said: "Science never solves a problem without creating 10 more."

Chapter 10

Magnetic Augmented Reality System of a Fox

Today, let's talk about foxes, mice, and the magnetic augmented reality system.

Let's start with an important piece of information. The main food of a fox is rodents, which make up 80–85% of its diet. To satisfy its hunger, a fox needs to catch and eat no less than two dozen mice a day.

Diogenes: I only eat mice in sour cream, which I catch in the refrigerator. In my fridge, there are mice arranged to look like sausages.

Ralph: "Excellent" sausage recipe: 100% mouse meat. And where did you buy those?

Diogenes: A friend of mine sent them to me, who was recently elected mayor of a small town in a northern country. They've opened a factory there for producing such sausages.

Reader: Ah, how good it is that our town's mayor is a human, not a cat. But I'd like to hear not about sausages but about foxes. How do they hunt and live?

Diogenes: Well, they hunt roughly like we, cats, but maybe a bit better. But their life is more complicated. They have to do more than just run after mice; they also have to jump.

Shadowless Squids: Stories of Physics in Nature
Vitalii Zablotskii and Tatyana Polyakova
Copyright © 2025 Jenny Stanford Publishing Pte. Ltd.
ISBN 978-981-5129-43-4 (Hardcover), 978-1-003-57062-2 (eBook)
www.jennystanford.com

Reader: Is a fox a cat or a dog?

Ralph: It barks, but it's not a dog.

Diogenes: It catches mice, but it's not a cat.

Reader: But how skillfully it catches them! It seems to sense them even under deep snow and dives headfirst into the snow without fail.

Ralph: I dare to assume that magnetic fields and even quantum effects play a role here. I recently read that animals know quantum mechanics.

Authors: As Niels Bohr said, "Anyone who is not shocked by quantum theory has not understood it." Have you seen quantum mice? Are they not shocked by quantum mechanics?

Diogenes: All mice are, in some sense, quantum objects: now they are in a certain vicinity of space, now they are not. Or, speaking the language of quantum mechanics, there is always a nonzero probability of a mouse appearing in a given area of space and within a certain time interval. As is known, quantum particles are capable of performing so-called tunneling transitions from one state to another, even when these states are separated by a large energy barrier. And we know that mice are great at digging tunnels and moving through them from one room to another.

Ralph: Yes, mice are indeed quantum. Schrödinger's cat has no sway over them.

Reader: Schrödinger's cat exists, but it only resides in quantum mechanics textbooks and popular science literature. However, scientific literature has reported that mice are capable of sensing the Earth's magnetic field through some quantum effects. There must be some exotic quantum compass inside mice.

Diogenes: Of course, they have a magnetic quantum compass. Otherwise, it's hard to explain how mice underground dig tunnels in the direction they need.

Reader: Aha... If mice underground or under the snow use the geomagnetic field for orientation, then the fox might know this and take it into account while hunting.

Ralph: Indeed, the fox jumps after the mouse very artistically, as if the direction of the magnetic field lines precisely shows her where the mouse is.

Diogenes: The fox is a good actress and an excellent hunter. She has a great sense of smell and hearing. Why would she need a seventh sense—the sense of the magnetic field?

Reader: Have you heard about the augmented reality system?

Diogenes: Yes, I've encountered something similar on a home gaming console. But tell me more.

Authors: It seems that foxes use the Earth's magnetic field to locate their prey. Apparently, this is the only animal capable of determining the direction and distance of its prey—a mouse hidden under a thick layer of snow—using the magnetic field. Researchers believe that foxes have an innate augmented reality system, allowing them to see what is invisible to the eyes.

Want to know what it looks like? Then imagine yourself as a fox in the middle of the silent snowy tundra. You're hungry, and somewhere under the thick layer of snow, a very appetizing mouse is hiding. But your eyes see only the glistening surface of the snow. Your nose is frozen and can't sense any smells, which, moreover, can't penetrate through the thickness of the snow. Your ears try to catch the slightest rustle made by the mouse. And the mouse might be just a few meters away, but you can't approach any closer because it will hear you and immediately hide in its burrow. What to do? Turn on the magnetic augmented reality system.

Turned on, ready... Now you see the force lines of the magnetic field and even see where the magnetic field is disturbed (has a deviation in direction) because of the mouse there. All clear, the target's coordinates are determined. One high jump and flight to the right spot on a parabolic trajectory... And the unsuspecting prey is in your teeth. This is quite a curious spectacle, easily found on YouTube. Fantasized enough? Now, step out of the fox's guise and return to your usual self, and we will try to explain to you the physics of this unusual hunt.

Just before the jump, the fox moves forward with very cautious small steps, carefully listening to the slightest sounds from the potential prey. So, the fox undoubtedly uses sound signals while preparing for the attack. However, scientists from the Czech Republic and Germany, after analyzing about 600 fox hunts for mice, concluded that in the vast majority of successful hunts, the head and body of the fox were oriented in the northeast direction just before the jump. In other words, 74% of successful jumps were made in the directions of the magnetic north or south pole, i.e., along the force line of the magnetic field. Interestingly, the success of such jumps along the force lines of the magnetic field did not depend on the depth of the snow cover, surrounding vegetation, weather (cloudiness and wind direction), time of day, or the geographical location of the hunting site. Thus, in the absence of any other source of information on the direction of the prey, the non-random orientation of the directions of successful fox jumps allows us to say that it is precisely the Earth's magnetic field that serves as a kind of navigator for the hunting fox.

Figure 10.1 My magnetic sense tells me: there's a mouse under the snow here.

This is not surprising, as mice also use the Earth's magnetic field for orientation under the snow. Scientists proved this by immediately placing a mouse caught in the forest into a special closed container equipped with two electromagnet coils. The coils were positioned so that they could create and change the direction of the magnetic field in the container to the opposite. Two minutes after capture, the mouse was transported in this container 40 m north of the capture site and then released. Observations were made on which direction the mouse would run over the next 4 min. It turned out that if there was no magnetic field in the container during transportation, the mouse ran south after being released, back to its capture site. However, if a magnetic field, opposite to the Earth's field, was created in the container during transportation, the mouse moved in the opposite direction from the capture site after being released.

So, mice use the Earth's magnetic field for their orientation, and foxes, knowing this and utilizing the geomagnetic field, precisely aim their jumps.

Ralph: It turns out interestingly: "magnetic" foxes hunt "magnetic" mice, who know quantum mechanics.

Authors: Yes, it may well be so. At least, it's a very fantastic idea, but it seems very plausible.

Chapter 11

Champions of Regeneration and Magnets

For thousands of years, doctors and wizards have sought living water capable of performing miracles. If, for instance, something unfortunate happens to a person's leg, and you have a flask of living water, it's enough to sprinkle a bit of this water on the injured leg, and, lo and behold—a new, perfectly healthy leg grows in its place. You might say all this is tales or fantasies. No, no! Nature has already created organisms that can regenerate (restore) not only damaged organs but the entire organism as a whole. These are planarians—small flatworms (1–3 cm) living in freshwater and saltwater bodies.

But what do planarians have in common with magnets? Here's the connection. But first a question. Can you separate the north pole from the south pole of a permanent magnet? Well, let's try. In our minds. Take a magnet in the form of a long parallelepiped, half of which, corresponding to the south pole, is painted blue, and the half corresponding to the north—in red. First, we cut it in half and instead of separate N and S poles, we get again two magnets: N-S and N-S with absolutely the same magnetic properties as the original magnet. Let's try to cut the resulting magnets. And again, failure, instead of separate poles, we have 4 magnets: N-S, N-S, N-S, and N-S (Fig. 11.1). Taking the next step, we get 8 identical magnets: N-S, N-S,

Shadowless Squids: Stories of Physics in Nature
Vitalii Zablotskii and Tatyana Polyakova
Copyright © 2025 Jenny Stanford Publishing Pte. Ltd.
ISBN 978-981-5129-43-4 (Hardcover), 978-1-003-57062-2 (eBook)
www.jennystanford.com

N-S, N-S, N-S, N-S, N-S, and N-S. And so it will continue ad infinitum. What about planarians? you might ask. Imagine a flatworm with a head on one end and a tail on the other. For example, let's depict it schematically like this: (: --------- <. You won't believe it, but if you cut a planarian in half, you get two planarians, and each will have a head and a tail: (: ---- < and (: ---- <. Cut these, each into two parts. You get four planarians: (: -- < (: -- < (: -- < (: -- <, as shown in Fig. 11.1. If you look closely at this figure, you'll notice that just as dividing a magnet into parts, and dividing a planarian, maintains a certain orientation of the resulting parts: the N-S direction in the new magnets and the head-tail direction in the little planarians.

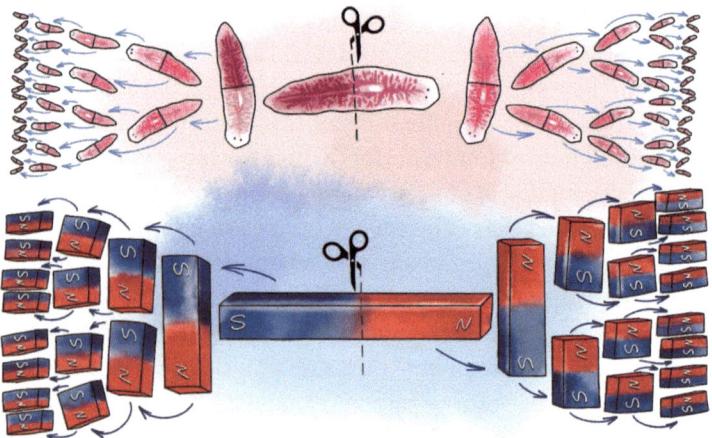

Figure 11.1   Will this really go on indefinitely? I can't believe it!

Can this division continue further? Yes. They can be cut into 200 pieces, and each of them will turn into a new individual approximately after 7 days. Scientists have determined the minimum fraction capable of regenerating into a full planarian as 1/279th of its body volume! Like a magnet cut into pieces, each piece of a planarian determines where the head and tail should be in the wound (the site of the cut) and transforms into an ordinary worm with one head and one tail. Thus, each piece of planarian somehow knows what a planarian should look like, remembering its left and right, as well as where the head and the tail are. Moreover, planarians combine these abilities with significant cognitive potential: they behave as the original worm did in similar conditions and can learn. For example, using a specially developed automatic learning device, it was shown that *D. japonica* planarians, having learned to recognize the rough texture of petri dishes, recognized it even two weeks after beheading followed by head regeneration. Normally, planarians need time to ensure the safety of a new environment before accepting the offered food, but previously trained new-headed worms did not have this delay, unlike regenerated individuals who had not encountered roughness. All this, at the very least, raises doubts that memory is a function exclusively of the brain with its synaptic connections between neurons.

Reader: You've used a good term: "new-headed individuals."

Authors: Yes, when applied to humans, it sounds strange. But perhaps in the future, it will sound more familiar.

But let's return to the theme of our story. One could say that planarians are champions in repairing their bodies.

Diogenes: Excuse me, but I'll interrupt you. It's known from biology that planarians are lower worms. But how can lower organisms have the most advanced regeneration system?

Authors: Apparently, planarians possess a special code (program) stored in all parts of the body, which is activated, possibly, by some yet unidentified cellular mechanisms. Here, one of the main mysteries is: (1) What or who "conducts" the development of a new organism from a small piece of the original planarian? (2) Where is the information about all the elements of the organism and the program for its full restoration stored? All this is astonishing and remains unclear, despite more than 200 years of studying planarians. To date, scientists have established that a particular type of stem cells—neoblasts, responsible for regeneration, are distributed throughout the planarian's body. But the most interesting and important thing is that, perhaps, DNA is not the only carrier of information and developmental program code in our world. Experiments with planarians show that there is also some mysterious bioelectric layer, coordinating the collective effort of cells in growing organs and bodies. Under the leadership of Professor M. Levin, scientists managed to crack the program coordinating the growth of a new planarian and create a planarian with two heads but no tail, and even a planarian with no head but two tails! Interestingly, planarians with two heads live a normal life, just like planarians with one head. It's very important that these planarians were not created through any genetic modifications!

Ralph: That's simply incredible! How did they manage that? Are they cool hackers?

Authors: No, they are biophysicists. They managed to interfere with the work of the planarian's bioelectric signaling system. First, they divided the planarian into three parts, and then immediately applied an electric field capable of changing the cellular membrane potential. And a miracle happened: from the part where the electric field was sufficiently large, a planarian with two heads began to develop. And from the part that was in an electric field of a different intensity, grew a planarian with no head but two tails.

Ralph: And if you cut the two-headed planarian in half, what will result from those halves?

Authors: M. Levin's experiments show that each of these halves also grows into a two-headed planarian! But let's return to magnets and planarians.

Since there is a certain analogy between dividing a magnet and a planarian (Fig. 11.1), the following question arises. How would planarian regeneration occur in a magnetic field?

Researchers report that the growth of the blastema—cells that transform into new parts of a planarian, slows down under the influence of constant magnetic fields from 100 to 400 µT. However, their growth accelerated in fields above 500 µT. They also discovered that the magnetic field altered the levels of ROS (reactive oxygen species)—they were lower than they would be under normal conditions for the blastema exposed to lower magnetic fields, and higher in planarians subjected to fields above 500 µT. Researchers have yet to explain the observed changes in planarian growth rates. But even more astonishing is the fact that the magnetic field can lead to the death of planarians. It turns out that if planarians are subjected to prolonged exposure to variable frequency-modulated magnetic fields, they fragment and die. And this is a new mystery.

Reader: It seems planarians have enough mysteries without a magnetic field. And the magnetic field only added new ones.

Authors: Yes, you're absolutely right. Nature has shown a lot on planarians, what it's capable of. And scientists are trying to find tools with which they could influence regeneration processes, which would allow them to understand the mechanisms and essence of these processes. Here, the magnetic field is one of such tools. Perhaps, when scientists understand how all this is arranged and organized in planarians, tales of living water will become reality.

Diogenes: Personally, the idea of someday meeting, e.g., a new-headed cat doesn't appeal to me. I should note that we, cats, have our own system of regenerating damaged tissues. You may have noticed that if you put your hand on a dozing cat, you can feel light vibrations in its body. This is its regeneration system at work, based on internal low-frequency vibrations of tissues. A cat purrs at a frequency of about 25–50 Hz, and such mechanical vibrations stimulate the processes of tissue healing and bone growth.

Ralph: And why specifically vibrations, and not some other physical process, are used for cat therapy? Or is it your secret?

Diogenes: I'll answer that question with the words of Nikola Tesla: "If you want to find the secrets of the Universe, think in terms of energy, frequency, and vibration."

Authors: Well, there's a hint for our scientists from the respected Diogenes. To discover the secret mechanisms of regeneration in planarians, look for their internal vibrations: mechanical or electromagnetic.

Chapter 12

A Touch of Mysticism: Memory, Hypnosis, and Wrinkles in Time

Memory is only possible against the backdrop of eternity...
—Cat Diogenes

If you're 20 years old or older, try to recall a vivid event from your childhood. For example, when you were 7 years old, your parents gave you a new bicycle you had long dreamed of. Have you remembered something bright from your childhood, how it was, and where it happened?

Now, the question. Are you sure you were in that place?

"A strange question," you might say. "Of course, I was there and remember how everything happened."

But we are now going to prove that you weren't there!

From biology, we know that the cells of the human body are constantly being renewed. And after seven years, a person consists entirely of new cells. Therefore, a well-known biologist joked, "If you haven't seen an acquaintance for seven years, you can refrain from shaking their hand upon meeting. This is already another person, as not a single cell from those that comprised them seven years ago remains."

Shadowless Squids: Stories of Physics in Nature
Vitalii Zablotskii and Tatyana Polyakova
Copyright © 2025 Jenny Stanford Publishing Pte. Ltd.
ISBN 978-981-5129-43-4 (Hardcover), 978-1-003-57062-2 (eBook)
www.jennystanford.com

Back to our question. So, if that significant event happened, say, 14 years ago, then from a physical or material perspective, you indeed weren't there, as all the cells of your body (30–40 trillion cells, $\approx 4 \cdot 10^{13}$) have been renewed twice in that time. In other words, not a single cell of your current body was there. And immediately, the question arises: where in the organism is human memory hidden? Where and how is information about past events recorded and stored in a person? Given the above, it's evident that memory is not in the cells. Either cell possesses the ability to retain information and transmit it to daughter cells upon division.

Let's try to find where memory could be stored in humans. We know that the human head is filled with electrical impulses generated by the activity of neurons in the brain. Perhaps these electrical impulses, constantly running through neural networks, are human memory. Of course, neurons, like any other cells, constantly die and are replaced by new neurons. But in the neural network, this process occurs gradually, so as not to disrupt the cooperative work of the electrical network, defining our everyday thoughts and actions and ensuring our long-term memory.

Can a person remember the details of events that occurred, say, 50 or 60 years ago? It is known that under hypnosis, people can recall and reproduce the minutest details of what happened to them quite a long time ago. Therefore, understanding the mechanism of hypnosis could partially lift the veil of mystery on our memory. However, the phenomenon of hypnosis is still not sufficiently studied, despite the fact that medical professionals have learned to induce it. Therefore, observing a hypnosis session gives rise to a feeling of mysticism in what is happening. Recall that hypnosis is a temporary state characterized by sharp focus of attention and high susceptibility to suggestion. Hypnosis is widely used to treat a number of diseases, e.g., anxiety, depression, and to alleviate chronic pain.

Not all people are equally susceptible to hypnosis. Previous research has shown that individuals highly susceptible to hypnosis have a strong (electrical) connection between two parts of the brain: the left dorsolateral prefrontal cortex and the dorsal anterior cingulate cortex.

So, it seems that the phenomena of hypnosis, memory, and electricity are somehow interconnected. In light of what we've discussed above, we can hypothesize that memory represents constantly living electrical impulses in the human head over time. And scientists indirectly confirm this.

For instance, sending electrical impulses to a specific part of the brain makes people more susceptible to hypnosis. Electrical impulses improve communication between the aforementioned brain parts responsible for susceptibility to hypnosis. To prove this, volunteers were administered 800 electrical impulses using electrodes placed on the scalp over 1.5 min. Magnetic fields were also used to stimulate nerve cells in the target brain tissues. Thus, 1.5 min of exposure to the brain with electrical and magnetic impulses allowed recalling long-forgotten events.

Think about it: why do electric current and magnetic field impulses applied to the head alter the state of the human brain and affect its memory? Is this mysticism or just simple physics plus complex biology? Perhaps Nikola Tesla was right when he said that electricity is us. And we are only beginning to realize this. Or is it still mysticism?

Ralph: All this looks interesting. But is our memory really as simply structured as the memory of a regular computer or phone?

Authors: Of course, our memory is far more complex and sophisticated. Unlike the magnetic or flash memory of a computer, the memory of living organisms is trainable, develops, and changes both for better and for worse. And most importantly, our memory is the basis of human emotions and creative activity. Computers are not yet capable of this.

Diogenes: I want to make a philosophical remark about the influence of electrical impulses on human memory and sensitivity to hypnosis. Socrates believed that "true knowledge" already exists in the soul of every person, and through skillfully posed questions, one can induce any individual to "remember" it. So maybe electrical and magnetic impulses play the role of those skillful questions, prompting people to recall events of long-gone days.

Authors: Thank you, Diogenes. In modern language, what Socrates talked about is called genetic memory, whose mechanism is shrouded in even greater darkness than the mechanism of ordinary memory. So, you've brought the discussion back to the realm of true secret knowledge and mysticism. But let's stay within the realm of science.

Diogenes: I want to philosophize a bit more about time and its perception by our brain. May I?

Authors: Yes, of course. Time, its perception, and memory are interconnected and fascinating subjects.

Diogenes: I believe there are waves of time. Why? Because waves are everywhere. For instance, sound waves can propagate in air, water, and any medium, while electromagnetic waves can even in vacuum. So, are time waves possible?

Let's try to answer this question. You can easily imagine sea waves. But how to imagine time without looking back into the past or peering into the future? The best image of time is a wide, flat river slowly carrying its waters to the sea. Indeed, here is an analogy: the flow of water in the river is the flow of time in our space. Not in vain did the ancient philosophers, trying to emphasize the flow of time and the impossibility of its repetition, say that you cannot enter the same river twice.

Imagine a river flowing across a plain. Its surface is very smooth, and the current is uniform and homogeneous. And in our Universe, time flows uniformly and homogeneously (if, of course, we are far from black holes!). Now, imagine a small wave on the river's surface, going against the current. Rest assured, such waves exist. When a wave spreads against the current, water particles lag behind their former neighbors. Conversely, a wave moving with the current carries water particles so that they outpace those with which they were recently nearby.

Let's transfer this picture of waves spreading on the river to the flow of time. So, waves can run through time both forward and backward. We're not discussing here the possible sources of such waves. Let's assume they exist. Then, people caught in a wave going against the flow of time fall slightly into the past. And people covered by a wave moving with the flow live in the future. If the wave is small, then the time shift is minor. Hence, the person does not explicitly notice it. They just navigate life better than others because they can anticipate some events, such as fashion trends, directions of social development, catastrophes, wars, and, of course, scientific discoveries. The latter relates to forecasters of the future and science fiction writers. Meanwhile, a person in the area of a reverse wave lives slightly in the past and, in the eyes of others, appears as a retrograde. Surely, you've met both types. I don't know what's better or worse. Probably, as they say, it's better to keep up with the times. But as you see, it doesn't depend on us. Time waves can catch up with us anywhere and anytime.

Reader: Interesting. So, you're saying there are wrinkles in time, right?

Diogenes: Yes, exactly. And possibly, the perception of "wrinkles" in time is related to the heartbeat and our sense of time.

All of us have internal clocks that count our internal time. And the heart rate (pulse) is one of our chronometers, reflecting the current rhythm of life. It is known that the average heart rate values in children significantly differ from the heart rate frequencies in adults. In other words, what is considered pathology in elderly patients is permissible for newborns.

As a child grows, the pulse rate gradually decreases and stops at the end of puberty (adolescent age) at a mark of 60–90 bpm.

It turns out that it seems to us as if time goes by slowly in childhood. And this may be related to the fact that children have a fast pulse. You can imagine it this way: one heartbeat equals one conditional second. And this means that in childhood, internal clocks count more seconds, minutes, and hours in duration for the same event than in adulthood.

Ralph: Dear Diogenes, thank you for the interesting hypothesis. But your philosophical thought suggests that children are on the waves of time running forward, and the elderly are on the waves of time running in the opposite direction. Right?

Diogenes: I'm not asserting that, but anything is possible. Surely you won't deny the fact that a child's thoughts are directed toward the future, while an elderly person's thoughts are in the past.

Authors: And we suggest ending the discussion on this philosophical note.

Chapter 13

Bees that Know Electricity and Magnetism

If the bee disappeared off the face of the Earth,
man would only have four years left to live.
—Albert Einstein

Einstein's words were not in vain. Bees are incredibly beneficial insects that pollinate plants. Besides, they give us honey. Humanity might survive without honey, but without plant pollination, there would be no harvest, and in about four years, Earth would face total famine. But this story isn't about that.

As you might have noticed, when talking about electrical and magnetic phenomena related to various living beings on our planet, we never miss the opportunity to highlight their use of other physical laws and phenomena. This time, talking about bees, we'll discuss their use of electricity and magnetism, optics, and thermodynamics. We'll also touch upon the remarkable mathematical abilities of bees. Bees rightfully claim the title of "the smartest and most useful."

Bees and electricity

During flight, bees accumulate a positive charge on their body due to friction with the air. This positive charge is precisely what they

Shadowless Squids: Stories of Physics in Nature
Vitalii Zablotskii and Tatyana Polyakova
Copyright © 2025 Jenny Stanford Publishing Pte. Ltd.
ISBN 978-981-5129-43-4 (Hardcover), 978-1-003-57062-2 (eBook)
www.jennystanford.com

need. As you might remember from one of our previous stories, the Earth's surface is negatively charged. Therefore, flowers, pollen, and nectar also have a negative charge. This physical fact—the mutual attraction of bodies charged with opposite charges—is what a bee uses to collect nectar and transfer pollen to another plant. Negatively charged pollen literally clings to the bee's body before it flies for nectar to the next flower.

Recent experiments have helped discover that bees and bumblebees can not only sense small electric fields but also extract useful information from the characteristics of these fields. After visiting one flower, a bee takes away part of its negative charge, changing the electric field around that flower. Another bee, approaching this flower, feels the changed field of the plant and understands that someone has already visited it, and there's no nectar left. See how efficiently and calculatingly everything is designed.

Incidentally, a worker bee cannot fly too far for nectar. Usually, they collect honey within a few kilometers from the hive. This is not because a bee cannot fly long distances, but because nectar serves as an energy source for the bee's flight, essentially fuel for its engine. During flight, it consumes the same nectar it collected. Therefore, if a bee refuels with nectar too far from the hive, it will use it all on the way back. Arriving at the hive without nectar, the bee queen or someone from her administration will strictly ask: where's the nectar you were supposed to collect? For this, they might be demoted to guard bees or nursemaids for little bees.

Although being a worker bee is not "all honey," as they say. Flying a lot and carrying nectar is hard work. A single bee brings about 40–50 mg of nectar to the hive at a time. Throughout the day, a bee makes about 10 trips to the field, thus bringing about 0.4–0.5 g of nectar.

Not every airplane pilot can make 10 flights a day. And bees often have to fly in bad weather, by instruments: bees have both a magnetic compass and an electric compass. Moreover, such work is dangerous. A bee might get caught in a charged spider web, which, as we already know, locally changes the electric field and disorients the bee. Here's an interesting example shedding light or, better said, adding a new mystery to the possible mechanisms of 3D orientation in bees.

From a beekeeper's tale

I've been breeding bees for over 20 years. And I thought I had a good understanding of their habits. But once I decided to repair a hive. To not disturb the bees, I started the repair when almost the entire bee family—15–20 thousand worker bees—had gone to collect nectar. Having finished the repairs before the return of the worker bees, who make up the bulk of the bee family, I put the hive back in its place. But this time I secured it one meter higher than it had been. Mathematically speaking, I placed the hive exactly in the same spot on the plane, with coordinates x_0 and y_0. But I increased the z-coordinate by about 1 m, i.e., it became $z = z_0 + 1$, because it seemed to me that the hive had been too low before. In the evening, the nectar-laden bees returned to the apiary. They circled for a long time over the spot where the hive had previously stood, i.e., in the area of space with the center coordinates x_0, y_0, and z_0. They couldn't find their hive! I didn't understand; they could see the hive but didn't want to or couldn't rise 1 m higher. Why? I can't answer. I don't know how much longer they would have searched, but I felt very sorry for them. After all, they had flown more than a kilometer fully loaded with nectar, and now they couldn't find their home. So, I lowered their native hive back to its place, and the bees rushed into it.

Reader: Does this story imply that bees have issues with 3D orientation?

Authors: No, we don't think so. This story suggests that bees have many orientation mechanisms, such as pre-flight dance instructions, besides those already listed. Which one is primary and at which stage of flight, we still don't know. But we'll talk about the dances later; that's more about mathematics than physics.

Bees and optics

Let's remember that light consists of electromagnetic waves. Humans can see within a very narrow range of electromagnetic waves, from wavelengths of 380 nm (which corresponds to violet) to 780 nm (red). Light with a wavelength shorter than that of violet light is called ultraviolet, and light with a wavelength longer than that of red light is called infrared light or thermal radiation.

Bees are known to have the ability to see in ultraviolet light. This not only expands the range of electromagnetic waves visible to bees and helps them see smaller objects than humans can, but their ability to see in the ultraviolet spectrum also aids in spatial orientation. Moreover, the unique structure of a bee's retina provides not only nearly a 360° view but also endows bees with the ability to distinguish polarized light from non-polarized light.

Ralph: A brief note from physics. Light is called linearly polarized if the vector of the electric field intensity of the electromagnetic wave oscillates along a certain direction in space. Light reflected off the surface of a dielectric is partially polarized.

Note that bees do not fly at night (which is why they do not need to see in the infrared spectrum), and during the day, a bee primarily sees sunlight reflected off the surfaces of plants and flowers (i.e., dielectrics), meaning it sees partially polarized light. One hypothesis suggests that bees can see or somehow detect the direction of light polarization. In simpler terms, a bee can see the direction of the electric field vector of the electromagnetic wave!

Bees and magnetism

Let's just say that a living bee has a magnetic moment. Yes, a bee is a flying magnet or a flying compass needle. While scientists search for a magnetic compass in birds and some terrestrial and marine animals, a bee already has a ready-made magnetic compass needle— its body. Undoubtedly, a bee can feel the direction of the Earth's magnetic field with its body. As you already know, in a magnetic field, a magnetic dipole moment (i.e., a small magnetic needle) always tends to turn itself to be parallel to the field direction. Speaking in the language of physics, one can say that a moment of force acts on the magnetic dipole moment from the magnetic field, which tries to turn the dipole moment parallel to the field direction. Undoubtedly, a bee knows this perfectly well and feels it with its body.

Bees and thermodynamics: The secret weapon of bees

Bees are industrious and beneficial insects. Hornets, on the other hand, are large predatory wasps that hunt meat to feed their larvae. This is enough reason for a war between them.

Hornets are undoubtedly not vegetarians. The powerful jaws of these insects allow them to easily kill and eat a caterpillar, bee, or even a small amphibian. For the growth and normal life of the colony, adult worker hornets are forced to continuously find new prey, kill it, and bring it back to the nest. These insects are simply programmed by nature for such behavior. For a hornet, a beehive represents an ideal hunting ground and a means to replenish food supplies. Firstly, bees gather in large numbers near the hive, making them easy to catch. Secondly, they are a valuable source of nutrients for hornet larvae. Finally, taking over a hive provides an opportunity to stock up on honey, which is especially important for large wasps, as small hornets mainly feed on sweet products.

Meat and honey are the main loot for these flying pirates. The preparation of hornets for attacking a hive is interesting. They send scout hornets, whose task is to pinpoint the exact location of the hive, as well as to assess the number of bees defending the hive and the amount of loot. Such a scout is well-armed and very dangerous. Hornets, like wasps, bees, and bumblebees, have a poisonous sting. However, a hornet kills bees using its powerful jaws, not its venom. An Asian hornet, e.g., is about 5 cm long, nearly three times the size of a worker bee. This gigantic predator can easily break the junction between the head and thorax of a bee with a single movement of its jaws, after which the bee can no longer move. In a minute, such a hornet can kill two to three dozen bees, and it only takes a squad of 30–40 predators a few hours to destroy a multi-thousand honeybee family.

It would seem that bees have no defense against this type of hornet. But as we already know, bees are not only proficient in mathematics but also in physics. Based on their scientific knowledge, bees have invented a secret weapon that allows them to kill scout hornets.

This weapon is a thermal bomb. A large number of bees literally envelop the predator, forming a huge ball around it up to 30 cm in diameter. "So what," you might ask. Bees are not warm-blooded, and their temperature is equal to the ambient temperature. Correct, but by forming this ball, bees actively work their wings. The heat generated from such muscular work heats the air inside the ball. Moreover, the heat generated from the movement of the wings is directed toward the center of the ball, i.e., to the attacking hornet.

Perhaps through evolution, bees learned that temperatures of 46–47 °C are lethal for the giant hornet. After an hour in such a ball, it dies, having managed to kill only a few of the bee defenders of the hive. Importantly, bees themselves can normally withstand temperatures up to 50°C, and those not caught by the jaws of the hornet dying from heat survive. It can be calculated that about five hundred bees are needed to create one such thermal bomb and kill one hornet. Having such living thermal bombs at their disposal, a bee family of 15–20 thousand worker bees can withstand an attack of 30–35 hornets.

Figure 13.1 We are flapping our wings harder. Our mechanical work is turning into heat that will kill this terrible predator.

In conclusion, let's briefly outline the physical principles underlying the bees' secret weapon. Firstly, the conversion of mechanical work into heat. Recall that the mechanical equivalent of heat was determined by Joule in 1843. In our case, the mechanical work done by bees through rapid movement of their wings leads to an increase in temperature around the hornet. Secondly, the bee ball acts as an almost adiabatic shell from which heat generated by the bees ($Q = 0$) cannot escape. In this case, according to the first law of thermodynamics ($Q = \Delta U + A$), all the mechanical work (A) of the bees goes toward increasing the internal energy of the system (hornet + bee ball) ΔU, i.e., increasing its temperature. And of course, when applying such a weapon, it is extremely important to know the limit temperatures of both the enemy and oneself.

Bees and mathematics

Bees are great mathematicians. Don't believe it? Then try to solve the following optimization math problem.

You need to construct a closed storage of maximum volume in the form of hexagonal prisms, using the minimum amount of building material. Calculate the geometric parameters of the lid that covers the cells.

Why exactly hexagonal prisms? Since ancient times, mathematicians have known that hexagonal packing is the densest. Therefore, bees did not even consider constructing cylindrical cells or cells of any other shape. The hexagonal form of the cells allows bees to make the honeycombs light, strong, and at the same time use the minimum amount of wax. This also enables them to use the available space most efficiently and store the maximum amount of honey. Bee honeycombs are rightly called an architectural masterpiece. So, what's the masterpiece if this was already known to ancient mathematicians?

It's all about optimizing the lid that covers the honeycomb. Bees close the cells using three rhombuses (equilateral quadrilaterals). The internal angles of the rhombuses, equal to 70.5° and 109.5°, represent the perfect mathematical solution to the problem of optimizing the shape of a lid consisting of three rhombuses. Note that solving this problem and obtaining the exact values of angles 70.5° and 109.5° requires knowledge of geometry and differential calculus at the level of a first-year student of a physics or mathematics faculty!

The method used in the construction of honeycombs is very precise and astonishing. Builder bees start constructing honeycombs simultaneously from 2–3 different points and build them in 2–3 rows. Thus, a large collective of bees, starting from different points, makes hexagons of equal sizes, connects them, and when finishing their work, bees meet exactly in the middle. It seems as if they have a ready-made project, and everyone knows and executes it precisely.

In this project and its implementation, the magnetic field plays a role again (since bees can perceive the magnitude and direction of magnetic fields). Honeybees use the Earth's magnetic field as a guide in constructing honeycombs. Probably, this is more accurate than leveling the honeycombs by eye, as humans do when crafting something at home. Interestingly, a swarm of bees building honeycombs without a foundation (an artificial base or template for building bee honeycombs) orients it in the same direction relative to the magnetic field as the honeycombs in the parent family were. Finally, the direction of honeycomb construction changes with the artificial alteration of the magnetic field.

Reader: You mentioned that hexagonal honeycombs are an example of dense packing and rational use of space. Does cellular communication also use this principle?

Figure 13.2   The flight map in the language of dance.

Authors: Cellular communication or communication network is a type of mobile radio communication based on a cellular network. The idea is that the total coverage area of the radio signal is divided into cells (honeycombs), defined by the coverage areas of individual base stations—transmitters and receivers (repeaters). The cells partially overlap and together form a network. On a flat surface, the coverage area of one repeater is a circle, so the network composed of them looks like a honeycomb with hexagonal cells. Recall: hexagonal packing is the densest of all possible packings. Thus, the space is optimally filled with radio signals. One could say that humans took this idea from bees. They have adopted other useful technical solutions from bees as well. However, for some reason, humans did not adopt the idea of dances from bees, which we will now discuss.

The libretto of the bee ballet

Bee dances present yet another mystery for scientists. Dance is one way bees communicate through a certain set of movements, which might seem chaotic to humans or, conversely, filled with mysterious meaning. Imagine a person witnessing classical ballet for the first time. The beautiful movements of the prima ballerina may seem musical and harmonious to them but meaningless. However, if you are an experienced theater-goer, you know the libretto, and every movement of the ballerina acquires a precisely defined meaning in your mind. The same goes for bee dances. We see a dancing bee but don't know what its complex movements mean. We are unaware of the libretto of the bee ballet and can only guess about the information encoded in the dance. Here's the most common hypothesis explaining the mystery of bee dance. By performing this dance, bees that have discovered nectar convey to other hive members information about the direction in which to fly to find the necessary plants, the distance to them, and the amount of nectar in them.

Thus, the dance is a schematic representation of the flight route to a nectar-rich location. The route is drawn, but where is the map?

With this dance program, we conclude the story of the amazing abilities of bees. Though one could talk about bees endlessly. But we see that not only our reader has dozed off but also his cat.

Reader: Cats sleep two-thirds of their lives. It's their lifestyle. And my cat listened to everything about bees, albeit with a rather gloomy look. I asked him what he was displeased with. And he said that you, in this story, didn't mention cats at all. At least that's how I translated his words from the cat language.

Authors: But we somehow didn't intend this book for cats. Please convey our apologies to your cat. In this story, we had no logical bridge to transition to cats, as we talked about hard-working bees. As known, a cat should not work. You said yourself that cats sleep two-thirds of their lives. However, we can end this story with a little plot taken from real life. The logic here is simple: bees, honey and wax, mice, and finally, cats.

Once at the market, we met our acquaintance, a beekeeper, when he was buying kittens. Knowing that he already has three cats at the apiary, we asked him why he needed so many cats. He replied that his three cats couldn't cope with the work.

And what do your cats do at the apiary? They do what cats are supposed to do—catch mice, he answered. I have a hundred wooden hives. So, these cheeky mice have gnawed holes in the hives and are stealing beeswax and honey from the bees. That's why I want to increase the number of cats to save the bees from the gluttonous mice.

Reader: My cat sends his thanks to you. And says he liked the story very much.

And you?

Chapter 14

Electromagnetic Trash

We live in a world entangled in wires. Wires are everywhere: above the ground, underground, and underwater. Wild animals have nowhere to hide from the electromagnetic fields created by the currents flowing through these wires. In addition, electromagnetic fields are emitted by transmitting antennas of cell towers, radars, mobile phones, and various operating electronic devices. Here, we're not discussing whether these fields affect humans because humans are the ones creating them. However, animals and plants are not to blame. They have managed perfectly well without all this electromagnetic trash for millions of years. And one of the main ecological questions is: do electromagnetic fields affect the animal world?

When you're out of town, take a look around. Undoubtedly, you will see power lines. The wires of high-voltage power lines create electromagnetic fields that extend tens of meters around them. It turns out that these fields have a noticeable effect on the state and behavior of bees. Biologists have shown that near power lines, bees experience stress and are less effective at pollinating plants.

As is well known, "If bees disappear, humanity will only have four years left." This phrase is attributed to Einstein. Regardless of

Shadowless Squids: Stories of Physics in Nature
Vitalii Zablotskii and Tatyana Polyakova
Copyright © 2025 Jenny Stanford Publishing Pte. Ltd.
ISBN 978-981-5129-43-4 (Hardcover), 978-1-003-57062-2 (eBook)
www.jennystanford.com

whether he said it or someone else did, without bees, a large part of plants, followed by animals and humans, will perish. And now, new research by biologists has demonstrated that electromagnetic fields from power lines interfere with bees performing their main function for nature—pollinating flowering plants. At a distance of 10 m from a power line, the magnitude of the magnetic field created by the electric current reaches up to 10 µT. To remind you, the magnitude of the Earth's magnetic field is about 40–50 µT. Thus, the magnetic disturbance from power lines is as much as 20–25% of the geomagnetic field's strength. Since it is considered proven that bees use the geomagnetic field for orientation (special receptors allow bees to feel natural electromagnetic fields), near high-voltage lines they get disoriented. This makes them aggressive, lethargic, and unproductive.

Initially, scientists conducted experiments in the laboratory. A hundred honeybees (*Apis mellifera*) were exposed to electromagnetic fields of varying intensity for 3 min. Increasing field intensity enhanced the bees' synthesis of heat shock proteins by up to 50%, which serves as an important indicator of cellular stress. Moreover, these bees showed decreased activity of genes involved in navigation and memory functions.

It was discovered that exposure to electromagnetic fields in honeybees disrupts their ability to forage, altering their magnetic navigation, learning, decision-making mechanisms, flight, and foraging activity, thereby impairing pollination activity. Bees visited California poppy flowers (*Eschscholzia californica*) growing near power lines three times less often than those plants located far from the lines.

Ralph: Wow! If we continue to entangle the Earth with wires, then plants will not be pollinated! It would be the real end of the world.

Diogenes: I think it's even scarier that electromagnetic trash changes gene expression, cognitive abilities, and animal behavior. Now I understand what happened to me last week during a mouse hunt. You know, my friends and I go out of town to hunt mice on weekends. Not because we lack food, but just for physical activity, for sports, for health. That time, I was hunting mice right under a power line and couldn't catch anything because something strange was happening in my head, as if it was spinning. And I even felt like I partially lost my sense of smell there.

Ralph: It's entirely possible. It's well known that during MRI, some patients experience dizziness—vertigo. And for some people under the influence of a magnetic field, their sense of left and right gets swapped.

Diogenes: Probably, cunning mice know these biological effects and deliberately dig their burrows under power lines, so cats and foxes can't catch them there.

Ralph: I don't doubt the abilities of mice. In the city, they're accustomed to rummaging through food trash, and outside the city, they've found electromagnetic trash to hide in.

Reader: I'm curious about what will happen next. Will animals and humans adapt to such an abundance of electromagnetic fields on our planet, or will there be a gradual mass extinction of species under the influence of intensively accumulating electromagnetic trash?

Authors: You've touched upon a global question of evolution. Currently, no one knows the answer to this question. We can only note that one of the mass extinctions of the animal world, which occurred on our planet about 42,000 years ago, scientists associate with a geomagnetic reversal—a change in the direction of the Earth's magnetic field.

Chapter 15

Collective Intelligence Living in Magnetic Homes

Scientists estimate that there are about 10^{15} ants living on our planet. This means that there are almost a million ants for every human being! And there are a thousand times more insects than ants. This is a huge army. Indeed, on three square kilometers of the Earth's surface, there are more insects than people in the entire world! The smallest ants measure just 0.7 mm, while the largest can reach up to 5 cm. There's a saying, "The ant is not big, but it digs mountains."

Ants, like honeybees, are social insects living under the motto "All for the good of the anthill," i.e., all for the good of society. These golden words are probably inscribed in the constitution of the ant state. The wise Jewish king Solomon, according to legend, understood the language of ants and could even talk to them. He said, "Go to the ant, you sluggard; consider its ways and be wise. It has no commander, no overseer or ruler, yet it stores its provisions in summer and gathers its food at harvest." What explains such obedience and industriousness in ants is still to be discovered by scientists, probably for many years. But what can we, in our time, learn from ants?

Oh, ants can do a lot! They can build homes that stand for a hundred years. And the ratio of the height of an anthill to its diameter in most cases follows the universal law of harmony and beauty of

Shadowless Squids: Stories of Physics in Nature
Vitalii Zablotskii and Tatyana Polyakova
Copyright © 2025 Jenny Stanford Publishing Pte. Ltd.
ISBN 978-981-5129-43-4 (Hardcover), 978-1-003-57062-2 (eBook)
www.jennystanford.com

nature—the golden ratio, i.e., corresponds to the ratio of 1:1.618. (1.618... is an irrational number, known as φ in mathematics). Interestingly, since ancient times, renowned painters have chosen the proportions of their canvases in this ratio. But how did ants come up with the idea of the number φ?

Perhaps they too love beauty and harmony. Ants collect various minerals, including micro-particles of gold, possibly for decorating their homes. They can find their way to food sources and back to the anthill, venturing several kilometers from home. Moreover, they can memorize the route and convey the exact route to other ants. In experiments where a treat was placed at the center of a maze, the scout ant quickly found the shortest route to the treat, returned, and communicated the route to its colleagues, who then followed that precisely defined path. It was found that the scout ant memorized the first 16 turns from the start of the maze and possibly counted steps from turn to turn, then passed this precise information to its comrades. How many turns when navigating a maze can you, dear reader, remember? The orientation methods of ants during their journeys remain a mystery to scientists. What serves as landmarks for them: stars, the Sun, smells, soil vibrations, sound signals, terrain relief, or electrical and magnetic fields? An ant can drag loads thousands of times heavier than its weight. In transporting granular materials, ants use a very clever mathematical algorithm. Overall, they are well-versed in mathematics and algorithms, which they use in construction, daily life, and in warfare. Ants domesticate and herd livestock. Ants can reach speeds of up to 7.6 cm/s. For a human, a comparable speed (measured in body lengths per second) would be almost 55 km/h. To cross water barriers, ants can form living bridges and rafts with their bodies. And ants can predict earthquakes and other natural disasters. For instance, before the severe Spitak earthquake in Armenia on December 7, 1988, ants, escaping the impending disaster despite the frost, emerged onto the surface and scurried about in the snow. But people did not understand these timely warnings.

But you might say, there's nothing for humans to learn from ants. We can do all this and even more.

Oh no, ants possess something humanity lacks. This something can be called "collective intelligence" or "superconsciousness," as opposed to our subconscious, which represents the dark corners of our soul. Perhaps it's science fiction, but it seems that ants are like walking neurons, which, similar to human brain neurons, exchange electrical signals, think, and make decisions. It is assumed that a large community of ants is capable of not only perceiving but somehow generating and transmitting vast amounts of information. In other words, there exists some form of communication between ants that allows them to quickly organize, make decisions, and take active actions, e.g., in response to external stimuli. By the way, we ourselves do not know who makes a particular decision in a given life situation. For example, you do not know how to act in a certain case and start considering options for solving the problem. Finally, you are enlightened and suddenly realize that you have found a solution and make a decision. But, recent experiments with functional MRI showed that in the human brain, the decision appears, as seen by the magnetic activity of neurons, fractions of a second before the person realizes that the decision has been made! And you think you decided to act in that situation exactly this way and not otherwise! Maybe someone else made the decision, and you were just informed about it by sending the corresponding signal to the brain's neurons? We'll leave it to philosophers to question where this decision came from and why you were presented with a fait accompli after it was made. Ants, apparently, do not face such questions. Having received a decision from the collective intelligence, they simply say "Yes, sir" and begin to execute it.

Thousands of scientific papers have been written about ants. This is not surprising because there is much to study about them. For instance, one significant scientific mystery is how ants can navigate using the Earth's magnetic field. Believe it or not, this is indeed necessary for them. Imagine you have shrunk to the size of an ant and are lost in a forest. How would you find your way home when every blade of grass looks like a giant tree, small uneven ground appears as mountains, and the sun or stars are hidden by tree canopies and clouds? Could you use smells? No, they are overwhelming from all directions, changing abruptly and chaotically from one plant to another. Thus, the only option left is the magnetic field, as it is omnipresent and its direction is constant both day and night. But to navigate by the magnetic field, you need a compass. It seems that ants, at least some species like those from the *genus Cataglyphis*, have one.

To test whether the sprinter ants from the *Cataglyphis genus*, also known as desert ants, navigate not only by the sun and landscape but also by the Earth's magnetic field, biologists conducted an interesting and simple experiment. When ants were returning home after foraging, scientists placed coils behind them that generated a magnetic field roughly equal in magnitude to the geomagnetic field but directed away from home. Surprisingly, upon activating the magnetic field, the ants began looking back as if they saw or sensed something, eventually changing their route toward the coil instead of heading home. As a result, the ants lost their way and couldn't return home. This unexpected discovery of a clear magnetic compass in *Cataglyphis* ants raises at least one more question: Do ants use the magnetic compass only for navigation, or do they also utilize it in ant nest construction?

Figure 15.1 Superant.

It turns out that the above-ground part of the ant nest, often called the dome, exhibits magnetic properties. This is surprising since the dome is made of dry grass and leaves with a small mixture of soil. Where then do the magnetic materials come from? Scientists examined dozens of domes in Siberia and the Altai Mountains. In the domes carefully washed of organic substances, they found microparticles of magnetic materials—iron oxides: magnetite (Fe_3O_4) and hematite (Fe_2O_3), as well as weakly magnetic materials such as ilmenite and garnet, and various metal sulfides. Ants find and use these magnetic minerals in constructing the dome. This explains the dome's increased magnetization compared to the surrounding soil.

Reader: Are you saying that ants build "magnetic" homes because they like living in a local magnetic field?

Authors: We don't know what ants love, but it can be assumed that they need an additional local magnetic field for orientation in the underground labyrinths of the ant palace.

Reader: You talked about the existence of a magnetic compass in ants. Where is it located, and how is it structured?

Authors: The magnetic compass is most likely located in the ants' antennae. Possibly, at the ends of the antennae, there are small chains of nano- or microparticles of iron oxides, which act as a tiny compass needle. From a physics perspective, positioning this needle at the ends of the antennae increases sensitivity to its slight rotation. But this is still just a hypothesis.

To detect magnetic nanoparticles in the ant's body, scientists used a highly sensitive quantum magnetometer—SQUID. It was found that in the bodies of worker ants *Neoponera inversa*, *Ectatomma brunneum*, and *Pheidole sp.*, magnetic nanoparticles (magnetite Fe_3O_4) were present.

Reader: So, what can we actually learn from ants? To navigate by the magnetic field?

Authors: No, we can do that with navigation devices. Predicting earthquakes and volcanic eruptions is something we can't do yet. Also, ants have efficient mathematical algorithms. For example, in computational mathematics, the so-called "ant algorithm"—ant colony optimization—is used. It's one of the effective polynomial algorithms for finding approximate solutions to the well-known traveling salesman problem, as well as solving similar optimal route finding problems on graphs.

Reader: And the last question. This "collective intelligence," as you said, requires fast and reliable communication between individual ants and the center. What can serve as signals for this?

Authors: Your question borders on the fantastical, and our answer will be in the same vein. It's known that ants "hear" with their feet and knees—they pick up mechanical vibrations in the ground. However, within the ant colony and beyond, a more reliable form of communication is necessary. If we entertain the notion that such a form of communication truly exists within ant communities, then the only candidates for communication signals would be electromagnetic field oscillations—electromagnetic waves. It's hard to believe that ants are capable of emitting electromagnetic waves of sufficient intensity to maintain communication over distances of even a few meters. Yet, entomologists from Stanford University (USA) have discovered that in ant societies, there are exceptional individuals that act as hubs—network servers or backbone nodes— through which message transmission occurs. They recorded 4,500 acts of communication among returning forager ants with each other and with guards protecting the entrances. It turned out that most carriers only contact a few "friends," while only a few communicate with hundreds of their cohabitants. So, anything is possible: perhaps individual ants are capable of emitting and receiving electromagnetic waves? After all, ants have antennae for a reason.

Reader: Oh, that's exactly what modern humans need—the ability for super communication—to be online 24/7 without the need for the internet or mobile phones. But we'll have to wait until evolution grants us antennae for receiving and emitting electromagnetic waves.

Authors: Interestingly, what does your cat think about this?

Reader's Cat: We, cats, know absolutely everything. We can even predict earthquakes, not from a dungeon, but from somewhere atop a cupboard. Evolution has long provided us with whiskers— antennae, but we prefer to remain individualists and communicate in person, not through some waves or another.

Chapter 16

Tsunami on the Beach

On a secluded island nestled among the vast ocean, an unusual story unfolded. Imagine the warm southern sun, a gentle blue sea, and a beach with golden sand where carefree tourists were enjoying their time. Among them was a family with a large red dog, basking in the serene atmosphere. Suddenly, the dog began to bark and howl piercingly, desperately trying to signal everyone to flee from the shore.

One of the vacationers remarked, "It senses a tsunami approaching," and started to hastily pack up.

"No, it's just the dog going crazy from the heat," another lounged on, dismissing the warning and burying his attention in his mobile phone. He explained, "Tsunamis are a series of long waves caused by movements in the Earth's crust underwater. They typically result from major earthquakes, underwater landslides, volcanic eruptions, or significant offshore explosions. Since we haven't received any earthquake alerts, there's no threat of a tsunami," he reasoned, confident in his scientific rationale.

"This is definitely not a tsunami," a third person added, believing his observation about the unchanged sea level was a scientific proof enough to dismiss the dog's alarming behavior.

Shadowless Squids: Stories of Physics in Nature
Vitalii Zablotskii and Tatyana Polyakova
Copyright © 2025 Jenny Stanford Publishing Pte. Ltd.
ISBN 978-981-5129-43-4 (Hardcover), 978-1-003-57062-2 (eBook)
www.jennystanford.com

The dog's owner countered, "My dog isn't crazy. She's never acted this way before. Something bad must be coming. Moreover, a drop in sea level isn't always a reliable tsunami indicator. Sometimes, the change can be too subtle or happen too quickly to suspect an impending danger."

The beachside debate about the potential of a tsunami went on for a few minutes without reaching any definitive conclusion. However, seeing the humans' indecision, the dog took matters into its own paws. It grabbed an infant from a stroller and dashed toward the nearest high ground. The startled vacationers chased after the dog to retrieve the child. Within 15 min, the dog and the infant were safely on higher ground, soon joined by the rest of the beachgoers. Just then, a giant wave crashed onto the beach, erasing all traces of their previous leisure spot.

Those who had made it to the hill realized the dog's decisive action had saved them from certain doom. The smart canine stood nonchalantly, accepting the gratitude of the people.

"But how did it know about the tsunami?" you might wonder. Animals possess an instinctive ability to sense impending danger, including tsunamis. They can detect environmental changes, such as subtle ground vibrations before an earthquake or peculiar water behaviors, and react accordingly. At least two physical phenomena that animals might sense when a tsunami approaches are sound waves and sensitivity to electromagnetic fields. Animals, more sensitive to low-frequency sound waves (infrasound) accompanying major earthquakes, can exhibit anxious behavior and seek safety. Some animals, including birds, fish, and insects, can sense changes in electromagnetic fields caused by earthquakes or water movement, aiding them in detecting an approaching tsunami. This instinctual behavior can lead them to migrate to safer areas, away from coastal zones, seeking higher ground.

Before the devastating 2004 tsunami, some animals in Yala National Park, Sri Lanka, became restless and moved away from the coast. In Chennai, India, biologists noticed many coastal birds abandoning their nests long before the tsunami hit. Before the 2011 tsunami, residents of Kohama Island observed many dogs and cats acting nervously, squealing, or attempting to flee to higher ground. In 2004, in Thailand's Khao Lak nature reserve, elephants used for tourist rides became agitated and refused to follow their trainers' commands well before the tsunami, allowing the reserve staff to evacuate them to safety.

Figure 16.1   The dog stole my child! Chase after it!

The main question arises: what signals do animals receive when a tsunami is approaching? It's known that animals can be sensitive to changes in the magnetic field. Recent scientific evidence has shown that an approaching tsunami generates a magnetic field around it. This happens because tsunamis move conducting seawater through Earth's magnetic field. Since seawater is salty and thus highly conductive, moving it through Earth's magnetic field induces an electric current, which in turn creates a magnetic field around the moving water. Dogs, like many other animals, might be able to detect these slight changes in the magnetic field, thus possessing the ability to predict a tsunami's approach.

Research has shown that the magnetic field generated by a tsunami can be detected about 10–15 min before sea-level changes, potentially improving the accuracy of tsunami predictions. This study confirms that the magnetic field created by a tsunami appears before sea-level changes and demonstrates that the magnitude of the magnetic field change can be used to estimate the wave height of the tsunami.

Diogenes remarked, "I always said animals should be trusted. They're smarter than humans in many ways."

Ralph added, "The wavelength of a tsunami ranges from tens to hundreds of kilometers. In deep ocean waters, tsunamis travel at high speeds, usually from 500–1000 km/h. As a tsunami approaches shallow coastal areas, its speed decreases due to interaction with the seabed. And the wave height? In the open ocean, near the area of origin, a tsunami's height ranges from 0.1 to 5 m. However, as it reaches shallow waters, the wave height can dramatically increase, reaching 10 m and even 50 m."

Chapter 17

The Price of Immortality

Immortality is a hope with which one should deceive oneself.
—Anatole France

"Would you like to become a *Turritopsis dohrnii* jellyfish?" someone asked a girl.

"No! Transform into a shapeless jelly and wander around the ocean? Why would I want that?"

"What if we tell you that this jellyfish is immortal. Scientists believe that some specimens of this species have been around since before the dinosaurs went extinct—about 66 million years ago—and still exist," someone explained.

"That changes things. I need to think about it."

"Consider this. The chance to see prehistoric dinosaurs and sea monsters, as well as to witness the distant future of our planet, is a good price."

"Oh. I'm almost convinced. Can I try on the jellyfish form?"

"Yes, we're turning on our experimental setup."

"Oh, I can see my long hair turning into tentacle-like filaments. And my body deforms into a transparent dome. Oh no, it seems I've become a jellyfish."

"But it's not so bad. I'm swimming in the ocean, looking around, enjoying the 3D freedom."

Shadowless Squids: Stories of Physics in Nature
Vitalii Zablotskii and Tatyana Polyakova
Copyright © 2025 Jenny Stanford Publishing Pte. Ltd.
ISBN 978-981-5129-43-4 (Hardcover), 978-1-003-57062-2 (eBook)
www.jennystanford.com

Figure 17.1   The elixir of immortality is beginning to take effect.

A voice from the fading laboratory said, "We forgot to warn you that the jellyfish's immortality is achieved through a repeating cycle from birth to old age and back again, without ever dying."

"That's no longer important," she thought. Suddenly, a predatory fish bit off a few of her tentacles. She felt no pain, but something inside her kicked in, like biological molecular motors starting up. Her tentacles began to retract into her body, and she slowly sank to the bottom. The last thought that flickered through her mind was, "The rebirth process has started, relax and rest on the sea floor for a while."

It's said that her body turned into a cluster of undifferentiated stem cells lying on the sea floor. Each cell contains all the information needed to construct a new whole organism, but only part of this information is actually used when the cell becomes differentiated. Here lies a heap of undifferentiated cells waiting for a signal to transform into a new organism.

And the signal came from somewhere. From the clump of cells, a hydroid (the basal stolon of a new hydroid) formed, from which a new polyp emerged. It's astonishing that jellyfish can transform directly into polyps, bypassing the fertilization and larval stages. This is even stranger than if a butterfly turned back into a caterpillar.

Of course, she didn't witness all these mysterious metamorphoses that happened to her, but when she transformed back into a young jellyfish with 8 tentacles, she quickly swam back to the laboratory.

The scientists understood without words that it was time to turn her back into a girl and activated their setup. Something buzzed inside her again. First, her eight tentacles turned into four arms and four legs, and she was terrified to see herself in the mirror. But then everything returned to normal, she dressed up, did her hair, applied makeup, and walked out onto the street. Only one thought troubled her: why did she so easily give up immortality and the chance to observe the evolutionary process on our planet?

Diogenes: As Horace said, "Not all of me shall die." So, don't worry.

Ralph: I have a few questions.

Where do "immortal" jellyfish live?

Turritopsis prefers warmer waters, though they've been spotted in colder regions too. They emerge in the *Caribbean Sea* (nutricula) and the *Mediterranean Sea* (dohrnii).

What do jellyfish eat?

Their diet consists of plankton, fish eggs, and small mollusks.

How big are they?

They're tiny, up to 4.5 mm in height and width. Young jellyfish have only eight tentacles and are 1 mm tall, while adults can have up to 90 tentacles.

Diogenes: What's the benefit to humans from studying these immortal jellyfish?

Authors: Transdifferentiation of cells may help scientists find new ways to repair or regenerate damaged tissues. What is the molecular mechanism that allows resetting the developmental information in all cells and moving to ontogenesis? The genome of *Turritopsis dohrnii* is being studied, and its decoding will be the first step toward finding the "immortality switch."

American physicist and Nobel laureate Richard Feynman said, "If someone tried to build a perpetual motion machine, they would be stopped by a physical law. Unlike that scenario, in biology, there's no law that mandates the finite life of each individual."

Diogenes and Ralph: We completely agree with that.

Chapter 18

How to Jump into a Flying Airplane

Did you know that worms use an electric field to jump onto bumblebees? *Caenorhabditis elegans* (*C. elegans*) are tiny worms, measuring just 1 mm in length and 50 μm in diameter, with one of the simplest nervous systems—considered simple because it consists of a small number of neurons. An adult hermaphroditic specimen is made up of 959 cells (males have 1031 cells) and has only 302 neurons. It seems incredible that such simple organisms could know anything about electric fields and derive benefit from them. After all, humans, with our 30 trillion (30,000,000,000,000) cells, took thousands of years of evolution and thousands of experiments to finally understand the electromagnetic field, thanks to Faraday (1831) and Maxwell (1855). And here we have the simplest worm, made up of a thousand cells, using electricity for movement.

No, *C. elegans* hasn't invented the electric motor or electric car. It goes even further. It directly uses the electric field to jump upward, e.g., to leap onto a bumblebee.

Diogenes: It's simply amazing. If we draw an analogy with airplane passengers, it paints an interesting picture. Just imagine: passengers using an electric field to jump onto a low-flying plane and calmly take their seats, just as these worms leap onto a bumblebee.

Shadowless Squids: Stories of Physics in Nature
Vitalii Zablotskii and Tatyana Polyakova
Copyright © 2025 Jenny Stanford Publishing Pte. Ltd.
ISBN 978-981-5129-43-4 (Hardcover), 978-1-003-57062-2 (eBook)
www.jennystanford.com

Ralph: It's known that in nature, smaller animals often hitch rides on larger ones to save their own energy during movement and to travel long distances. For instance, tardigrades can travel on snails; sea anemones hitch rides on the backs of hermit crabs; remoras—fish ranging from 30–90 cm in length, with suckers on the top of their heads, attach themselves to the undersides of passing rays or sharks.

Researchers previously found *C. elegans* on the bodies of flying insects but thought the worms climbed onto their carriers when they landed on the ground. However, scientists have now learned that worms can jump onto their "carriers" while the insect is in flight. It is known that *C. elegans* nematodes are transported through electric fields for phoretic attachment to insects. It was discovered that *C. elegans*, located at the bottom of a Petri dish, move to the underside of the lid directly through the air under the influence of an electrostatic field generated by an electric charge placed on the lid. Moreover, the worms jumped upward regardless of whether the charge on the lid was positive or negative. This means the worms don't have their own charge; their attraction to the Petri dish lid is due to the charge induced on their surface by the electric field.

Ralph: I understand. Worms are attracted to the charged lid in the same way small pieces of paper are attracted to an electrified ebonite stick or comb.

Diogenes: Yes, this experiment can be conducted at home by the reader.

Scientists have established that the average flight speed of the worms is 0.86 m/s (close to a pedestrian's speed) and the worms take off under the effect of an electric field exceeding $E = 200$ kV/m, with their speed increasing as the field strength increases. Interestingly, these "smart" worms also perform group jumps. Scientists observed a chain of 100 worms jumping onto the lid of a Petri dish. To check whether the worms could jump onto bumblebees, scientists charged a *Bombus terrestris* (*B. terrestris*) bumblebee with an electric charge $q = 806$ pC, within the range of normal electric charges of bumblebees and corresponding to the charges of flying insects observed in nature. When near such an insect, the worms immediately lifted vertically and began jumping onto the bumblebee's abdomen, sometimes even piling on top of each other and setting off in flight as a group of 80 "passengers" at once.

These observations led to the conclusion that *C. elegans* can attach to the *B. terrestris* bumblebee, which gets charged through friction with flower pollen.

Diogenes: This demonstrates the possibility of electrical interactions between different species of living beings.

Ralph: Yes, some pollinating animals, such as bees and hummingbirds, have a natural electric charge. This allows plant pollen to be attracted to the pollinator.

Reader: If we let our imagination run wild, we might soon hear an announcement at the airport: "Attention passengers, the last call for boarding. In 5 min, the electric field will be switched on for passengers of flight F2100. The low pass of aircraft F2100 at a speed of 240 km/h over runway No. 2 is expected."

Diogenes: You're looking into the future but let me tell you something interesting from the past. Scientists froze and revived roundworms that had been frozen in Siberian permafrost for 46,000 years. Moreover, in the laboratory, scientists cultivated more than 100 generations of these worms. Comparing the genomes of these worms with *C. elegans* revealed many overlapping genes related to survival mechanisms in harsh conditions. This is strange, as *C. elegans* typically lives in temperate climates.

Ralph: Wow! That means they were in a state of cryptobiosis for 46,000 years. Perhaps this is possible precisely because they are the simplest, just a thousand cells.

In conclusion, it must be said that *C. Elegans* are indeed primitive living beings. And the fact that they use an electric field for boarding does not speak to their intellectual abilities. It is nature, all-knowing in physics and biology, unlike us, that designed these worms to move great distances.

Diogenes: But why?

Ralph: "Because nature wanted it that way, and why is not our concern," as the song goes.

Chapter 19

Who Flies Without an Engine

Ralph: Hello Diogenes. Why are you lying there in a pensive pose? What are you thinking about?

Diogenes: I'm thinking about how creatures like you consume too much energy and heat up the Earth with your warmth. It won't be long before we're facing global warming.

Ralph: And are you an exception? Do you not warm the Earth?

Diogenes: We, cats, are very economical creatures. We don't make a lot of unnecessary movements. For example, we don't run around the yard for no reason, nor do we bark at the Moon.

Ralph: Oh, you cats sleep for two-thirds of your life!

Diogenes: Yes, but we don't just sleep. We doze and think at the same time. Today, I dreamt that I grew wings and was able to fulfill my dream.

Ralph: Which is?

Diogenes: I was soaring in the air and catching bats.

Ralph: A cat with wings, gliding under the clouds. I can't imagine it. A horse with wings—that everyone knows. That's Pegasus—the inspiration for poets. Though from a physics point of view, for a horse to fly, it would need very large wings. I think about 6 or 7 m long. But no, I'm more likely to see a horse swimming on its back than a cat with wings chasing after bats.

Shadowless Squids: Stories of Physics in Nature
Vitalii Zablotskii and Tatyana Polyakova
Copyright © 2025 Jenny Stanford Publishing Pte. Ltd.
ISBN 978-981-5129-43-4 (Hardcover), 978-1-003-57062-2 (eBook)
www.jennystanford.com

Diogenes: I see no reason why cats shouldn't be able to fly. From a physics standpoint, a cat's flight would be a more energetically efficient mode of transportation than running and jumping. I'll tell you about how condors fly, and you'll understand why I'm thinking about energy conservation.

Ralph: That's interesting. Tell me more.

Diogenes: Condors fly practically without a motor. By tracking the behavior of eight Andean condors, ornithologists discovered that active flight (flight with wing flaps) accounts for about 1% of their total flight time, which is a record low for birds. Individual birds were able to cover up to 172 km in 5 h without flapping their wings once. Why is that?

The reason is that for large birds, each wing flap is very costly in terms of energy. This fully applies to the Andean condor (*Vultur gryphus*), which inhabits the western part of South America. The wingspan of this huge bird can reach 3 m, and it can weigh up to 15 kg.

Ralph: Diogenes, you probably weigh as much by now. That means, for you to glide like condors, you'd also need wings with a span of 3 m! That would be an incredible sight—more like a dragon than a cat.

Diogenes: Condors are very smart and therefore prefer a passive form of flight. They can glide for long periods, like gliders, using rising air currents—thermal or those caused by the terrain. Nevertheless, even these giants sometimes have to make an effort and flap their wings. Scientists wanted to find out under what conditions this happens.

They attached sensors to eight condors that recorded every wing flap, altitude above sea level, and location. The total weight of this equipment did not exceed 1% of the bird's total weight. After 10days, the sensors detached and fell to the ground, where they were collected by researchers.

In total, the team managed to record data on 235 h of condor flights. The birds spent most of their time traveling between roosting and feeding sites, as well as patrolling areas where herds of livestock were concentrated—a source of carrion.

Observations of these condors showed that they rarely flapped their wings, spending just 1.3% of their total flight time doing so.

This is the lowest figure among all birds: for comparison, white storks (*Ciconia ciconia*) spend 17% of their migration time in active flight. For wandering albatrosses (*Diomedia exulans*), this figure ranges from 1.2 to 14.5%.

Figure 19.1 The winged cat is hunting for bats.

During long journeys, active flight for condors accounted for only 0.8% of the time, and during short ones—8.6%. The birds most often had to resort to active flight at low altitudes. In 75% of cases, this occurred during takeoff from the surface. Considering the high energy cost of wing flaps, it can be assumed that condors must carefully choose their landing spots to avoid excessive energy expenditure. Another situation in which condors flapped their wings was when transitioning from one rising air current to another.

Ralph: Yes, condors are very economical birds that know the laws of aerodynamics. And winged cats would probably be even more economical. But have you noticed that condors only glide freely at high altitudes, where bats no longer fly?

Diogenes: Keep your jokes to yourself. Some species of bats are capable of reaching altitudes over 2,000 m above sea level. For example, the Brazilian free-tailed bat (*Nyctinomops laticaudatus*) was spotted at an altitude of about 2,200 m in the Andes. They can reach such heights by using aerodynamic currents and thermal updrafts.

Ralph: Sorry, my mistake. You know, mice aren't my specialty. You recently defended your doctoral thesis on mice.

Diogenes: The findings from the condor flight studies are of interest to engineers developing unmanned aerial vehicles. And us, cats, evolution will eventually give small but effective wings, allowing me to fulfill my dream.

Ralph: Maybe evolution will grant cats antigravity instead of wings? Then you could fly silently and without a motor. Better than condors.

Diogenes: That's not a bad idea. I'll try to realize it in one of my next dreams.

Chapter 20

Spider on a Flying Carpet

Let us start this story with a riddle.

Riddle

These "adorable" creatures do not run on the ground, do not fly through the air, and do not swim in water or underwater. Do you know who they are?

No. Where do they live then, and how do they move?

They live in every corner (literally) and in all places on our planet. And they mainly move along special symmetrical suspended roads that they build themselves.

Still haven't guessed who we're talking about? Then here are a few more hints in the form of questions and answers. You ask, and we answer.

How do they hunt?

They hunt without leaving home.

Do you mean to say that their prey comes to them at home?

Yes, that's exactly what happens.

Are you suggesting that aliens from other galaxies have settled on our planet?

No. Although there is some resemblance to aliens.

Well, in that case, I can't figure out who or what it is.

Shadowless Squids: Stories of Physics in Nature
Vitalii Zablotskii and Tatyana Polyakova
Copyright © 2025 Jenny Stanford Publishing Pte. Ltd.
ISBN 978-981-5129-43-4 (Hardcover), 978-1-003-57062-2 (eBook)
www.jennystanford.com

Alright, one more hint. These creatures are the most gluttonous on Earth. Guessed who?

It's humans!

Wrong! Humans are second in gluttony. First place goes to spiders!

Indeed, they are the most gluttonous on our planet.

Scientists have somehow counted all the spiders on Earth. It turns out that about 45,000 species of spiders inhabit our planet (and those are just the ones scientists know about). But the most terrifying numbers are these. On average, there are about a thousand spiders per square meter of the Earth's surface!!! Quickly check how many spiders per square meter live in your apartment. Think zero? It's good if it's really so. But you might not have noticed the little spiders. And scientists have counted them too. Moreover, according to their calculations, the total mass of all spiders living on Earth is 25 million tons! So, how much does a spider eat?

Scientists pondered this question and found out that all spiders eat 400–800 million tons of food a year. For comparison, the amount of meat and fish that humans eat per year is about 400 million tons. We don't know the exact number of spiders on Earth, so we can't calculate how much prey one spider eats per year. But knowing the total mass of all spiders and the mass of prey they consume per year, we can introduce and calculate a new quantity—the specific gluttony of spiders. Analogous to specific heat capacity, we define the specific gluttony of spiders (voracity) as the ratio of the mass of food they consume to the mass of spiders, voracity = (800 million tons of food/25 million tons of spiders) = 32. This means that 1 kg of spider (or better said: a spider weighing 1 kg) eats 32 kg of prey in one year.

Reader: Interesting, I imagine a spider weighing 1 kg with 32 kg of prey it needs to eat. Can you calculate the specific gluttony of a human?

Authors: Nothing easier. First, we calculate the total mass of the human population living on Earth. Let's say there are now 7 billion people living on our planet. Take the average weight of a human as approximately 70 kg. Then the mass of all humanity equals 7,000,000,000 × 70 = 490 million tons. Finally, the specific gluttony of an average human equals voracity = (400 million tons of meat and fish/490 million tons of humans) = 0.82.

Reader: Wow! A spider is 40 times more gluttonous than a human! No wonder my cat is afraid of spiders.

Authors: Well, as they say, facts are stubborn things. And your cat is undoubtedly a major biophysicist.

Reader: My cat is sitting here listening to all this. And he has a question too: what do spiders eat?

Authors: Ah, your cat becomes more talkative. We remember how in one of the previous stories, your cat didn't want to share his knowledge with us. Tell him not to worry. Spiders don't eat cats. Spiders' diet mainly consists of insects. Some large species of spiders, especially tropical ones, consume frogs, snakes, fish, and lizards.

The gluttony of spiders depends on their habitat. Forest and meadow spiders consume more food than, e.g., desert and Arctic dwelling ones. Most likely, this is due to the different amounts of insects in these regions.

Reader: Are you saying that if the Arctic were teeming with flies and other insects, spiders would just sit among the snow and ice?

Authors: Well, why not? There's the polar bear, so why not have a polar spider if there's prey for it.

Reader: I imagined a gigantic white spider on an ice floe, catching northern birds with its charged web. And now I feel uneasy. Let's talk about something more cheerful.

Authors: Alright. Let's go on a journey on a flying carpet.

The electric flying carpet

You might find this surprising, but spiders can fly great distances, hundreds of kilometers, and they do this using their intellect and the Earth's electric field. Remember, in the story ("We Live on the Charged Sphere"), we detailed the reasons behind the Earth's electric field. To remind you, the Earth's electric field (E) strength is approximately 130 V/m, and its surface charge is negative. Moreover, the field strength increases before a storm. Spiders, although they haven't read our book, are well aware of this and use the Earth's electric field to fly on their own version of a flying carpet. You probably had fairy tales about flying carpets read to you in your childhood, while spiders sat quietly in a dark corner listening. And they didn't just listen; they took these stories to heart, intending to put them into practice.

You know what spiders do in their free time. Correct, they weave webs—a complex geometric pattern made of the thinnest, strongest, and most elastic threads. But what you may not know is that spiders, in creating their webs, charge them with a negative electric charge. Why?

It's known that like electric charges repel each other. If the web is charged with charges of the same sign, it won't tangle! Imagine holding a vast, delicate net in your hands. It would instantly tangle in such a way that you could never untangle it. That's why the spider charges it with an electric charge as it's made. Due to the electric repulsion between different sections of the web, it remains spread out and even taut. But we'll talk about its tension later. For now, let's focus on the web's electrical properties. The spider weaves silk threads, which it releases from various spinnerets inside its body, and then charges them by rubbing them against tiny combs on its hind legs.

So, the web is negatively charged, and its total charge equals q. As we know, the Earth's surface is also negatively charged. This means the electric force, $F = qE$, repels the web from the surface and even lifts it into the air if this force exceeds gravity, i.e., if $qE > mg$ (here m is the mass of the spider along with the web, and g is the acceleration due to gravity). So, for a long journey, the spider just needs to sit on its charged web and wait for the right flying weather. However, "flying weather" for it is the exact opposite of what it means for airplane pilots. For the spider-traveler, "flying weather" means the onset of a storm, when the electric field between the Earth's surface and storm clouds increases. During a storm, the electric field strength at the Earth's surface can rise to tens of thousands of volts per meter. Scientists from the University of Bristol showed that spiders can sense the Earth's electric field and use it to launch themselves and their webs into the air. When the electric field reaches the calculated value, $qE = mg$, the spider detaches the web and takes off. This is somewhat analogous to the concept of "lift-off speed" for an airplane from the runway, which is determined when the lift force of the plane's wing (proportional to speed) equals the gravitational force acting on the plane. Probably, in the flight manual issued for spider pilots, there is a concept of "lift-off field."

One could say that spiders have realized humanity's ancient dream of flying on a flying carpet. Using this efficient mode of transport, spiders can escape from their predators or competitors. They can even venture to new lands where food is abundant and new acquaintances await. It was observed that spiders could ascend to heights of up to two and a half miles in the air and travel 1000 miles away from the shore into the sea.

Reader: And how did scientists prove that spiders can sense the electric field and fly using electric repulsion from the surface?

Authors: First and foremost, researchers showed that spiders can sense electric fields. They placed spiders at the center of a polycarbonate box. Then they generated an electric field between the ceiling and the floor, roughly equivalent to that at the Earth's surface. It was observed that the field (electric force) moved the tiny sensory hairs—trichobothria—on the spiders' legs. Simply put, the hairs on the spiders' legs moved when the electric field was turned on, similar to how a person's hair stands on end when connected to a school Van de Graaff generator. As the electric field increased, the spiders performed a series of movements that resembled tiptoeing, i.e., they stood on the tips of their legs and lifted their bodies. Some of the lightest spiders even took off, although they were in a closed space where there are no air currents. When researchers turned off the electric field, the flying spiders fell to the bottom of the box.

The perfect hunting machine

Previously, we believed that only humans were capable of inventing various mechanisms to enhance their strength (like levers or hydraulic presses) and machines to produce energy and increase power. But spiders have proven to be no less ingenious in this regard. They have designed a device to externally amplify mechanical energy using their webs. The spider *Hyptiotes cavatus* utilizes the elastic energy stored in the threads of its web to actively ensnare prey from a distance. *Hyptiotes* tensions its web by pulling each anchor thread through several cycles of limb movement, thereby storing elastic energy in the threads. The result is akin to a tightly drawn bowstring. When an insect hits the web, the spider releases the anchor thread. As a result, under the action of elasticity and stored energy, both the spider and the web leap forward 2–3 cm with a peak acceleration of up to 772.85 m/s^2 (note that this acceleration is about 80 times greater than $g = 9.8$ m/s^2!) directly toward the prey, while the jerks caused by the spider's sudden stop lead to the prey being wrapped in the web from all sides. Using the web as an external "mechanism" for storing and releasing elastic energy accumulated over several cycles of muscle contraction offers significant advantages over the forces that could be achieved solely by the anatomical muscles of the *Hyptiotes* spider. The mechanism for amplifying elastic power used by this spider bears a remarkable resemblance to mechanisms created by humans, such as catapults and ballistae.

However, the web is not just any net, even if it is electrical. Its threads are also coated with a special adhesive that not only traps insects caught in the web but, thanks to its electrostatic properties, also causes the web threads to attract nearby flying insects. This is because, during flight and friction with the air, insects accumulate a static electric charge. Airplanes also accumulate charge during flight. You may have noticed that airplanes are grounded (connected by a special wire to the ground) as soon as they taxi to the jet bridge. Scientists have discovered that the sticky and charged spirals of the web alter the magnitude and direction of the Earth's electrostatic field a few millimeters away from the web. Many insects, e.g., bees, use electric magnetic fields for navigation and orientation during flight. Thus, for insects, the web is not just a mechanical net but also a clever trap with a disorienting electric field. Again, by analogy with airplane navigation, one could say that for insects using the Earth's electric field for orientation, the web locally changes the GPS signal.

If you still doubt the intellectual capabilities of spiders, let us add another stroke to their portrait. It is known that during hunting, a spider sits at the top part of the web to use the Earth's gravity to quickly move toward its caught prey. Of course, from a physics standpoint, it's obvious: running downhill is easier than uphill. Moreover, a thin, tightly stretched web in the shape of a disk acts as an ultra-sensitive acoustic antenna, capturing sound waves with sensitivity greater than the acoustic sensitivity of all previously known eardrums of other animals. Sensing the movement of the web's antenna threads, the spider remotely detects and locates the source of incoming air acoustic waves, e.g., emitted by approaching prey or enemies.

And a final example of how smart spiders are. One African spider spun its web between the horns of an antelope. Why? The reason is quite simple. When the antelope runs, it catches flying insects with this web just as fishermen catch fish with a net pulled by a fishing vessel. This spider is simply a genius.

Interestingly, the fastest, most agile, and precise living creatures on Earth turned out to be crab spiders, capable of turning 360° and delivering a precise strike to their prey in just an eighth of a second. Scientists came to this conclusion after observing these arthropods catching crickets and other insects with a high-speed camera.

Figure 20.1 For a spider, the best flying weather is a thunderstorm.

They're beneficial: Spider venom protein blocks "death waves" during a heart attack

It turns out that a medicine made from the venom of the deadliest spiders can help save human lives in the face of such dire human ailments as heart attacks and strokes. Scientists have developed a medicine based on the protein Hi1a, found in the venom of the Fraser Island funnel-web spider. This protein can be used as a drug that prevents the death of heart cells resulting from a heart attack. The Hi1a protein blocks certain ion channels, thereby interrupting the death signal to heart cells if they do not receive enough oxygen, e.g., during a heart attack or when the heart is removed for transplantation.

Spider farms

Sounds strange, doesn't it? But they could appear in the near future. The reason lies in the unique properties of spider silk and the prospects for its use as high-quality air filters. Indeed, the nanothreads woven by spiders, thanks to their electrical properties, have the ability to attract and retain charged particles, which fill the air in modern cities. To have such perfect nanofilters on an industrial scale, spider farms are needed, preferably of large and venomous spiders, to have a dual benefit: silk for filters and venom for medicines.

Epilogue

As we already know, spiders know and utilize three types of forces: electrical, elastic, and gravitational. Moreover, for hunting, they create simple mechanisms that serve to increase their strength and power. And in their free time, in weather not suitable for us, spiders embark on long journeys on flying carpets. Additionally, they bring certain benefits—based on their venom, scientists develop medicines for deadly diseases. So, maybe we shouldn't kill them at the first encounter? Remember, when you kill a spider, you reduce the spider population in your home. But at the same time, the remaining spiders become smarter and more cunning, since you are essentially removing the genes of the less intelligent spiders from their population.

Diogenes: I also think that spiders should not be killed. Maybe they are not really spiders at all. An ancient Greek myth from Ovid's "Metamorphoses" tells of the contest between the weaver Arachne and Athena, the goddess of weaving. After suffering defeat in this contest, Athena transformed the weaver Arachne into a spider.

Chapter 21

The Marriage of the Chiral Goat

Let's start with a joke. "John walked up to the mirror and raised his right hand. And then, to his horror, he discovered that his reflection in the mirror raised its left hand! 'A violation of chirality,' he thought in terror."

And what is chirality? Chirality is the lack of symmetry between the right and left sides. If an object's reflection in a perfect flat mirror differs from the object itself, then the object possesses chirality. In other words, chirality is the property of an object not to coincide with its mirror image through translations and rotations. In physics, chirality is a property of particles or fields that manifests in the difference between right and left.

In biology, unlike inanimate matter, living matter possesses homochirality (chiral purity). With rare exceptions, all proteins are made up of amino acids with left-handed chirality, and the residues of sugars deoxyribose and ribose in DNA and RNA molecules in all organisms have right-handed chirality. The mechanism of the evolutionary emergence of the chiral purity of proteins and nucleic acids remains unclear. The chirality of molecules is very important for life. For example, the well-known vitamin C—ascorbic acid—can exist in two configurations: left and right. However, only L-ascorbic acid with the left configuration of the molecule is vitamin C, i.e., beneficial for the body.

Shadowless Squids: Stories of Physics in Nature
Vitalii Zablotskii and Tatyana Polyakova
Copyright © 2025 Jenny Stanford Publishing Pte. Ltd.
ISBN 978-981-5129-43-4 (Hardcover), 978-1-003-57062-2 (eBook)
www.jennystanford.com

Reader: I have poor spatial imagination, so I don't understand your geometric definitions of chirality. Could you explain the property of chirality with a simple example?

Authors: Let's ask Ralph to explain to us in simple terms what chirality is.

Ralph: With pleasure. I'll tell you an old story in which chirality determined the fate of an individual. So, here's the story.

The chiral goat

This story was told to me by my grandfather when I was just a little pup. Ralph scratched behind his ear and continued. Long ago, he herded goats in the mountains. Being a very educated and attentive dog, he noticed a young goat whose horns were twisted differently than the other goats. If the horns of typical representatives of this goat breed are twisted in opposite directions: one clockwise and the other counterclockwise, then this goat, named MeX, had both horns twisted clockwise, presenting a right-hand screw, like a DNA molecule. However, this chirality (the chiral purity of his horns) did not manifest in his behavior, and he had a calm, but firm character. And all was well, but time passed, and one day MeX decided to marry a young and pretty goat Desiree from the neighboring herd. Here, unexpectedly, a problem arose. The elders of the neighboring goat tribe did not consent to their marriage due to MeX's obvious chirality. And no scientific arguments could persuade the leaders and scholars of this herd that chirality is not an obstacle to marriage. "Why do we need many chiral goats in the herd?" they reasoned.

But MeX's love for Desiree was so strong that he decided to undertake a scientific experiment that could help him correct the homochirality of his horns and win Desiree as his wife. He had heard somewhere that the chirality (direction of twist) of the first biological molecules was initiated by circularly polarized light from a distant star that had gone supernova. Indeed, he thought, light polarized in a left circle could help me grow a new horn, in the form of a left spiral. "I shall saw off one horn with the right twist and grow a new one under the influence of light polarized in a left circle," decided MeX.

At MeX's request, the shepherd sawed off one of his horns. And on the very first starry night, MeX went to the top of a high mountain to expose his head to the left circularly polarized light coming from a distant star. And, of course, you won't believe it (...and rightly so), but it all worked out for him. After some time, a new horn with a left twist grew in the place of the sawed-off horn. MeX was very pleased because he became a non-chiral goat with horns twisted in opposite directions. The elders of the neighboring herd consented to his marriage with Desiree, and soon the wedding took place. Here's your happy fairy tale ending. But the most interesting part is yet to come. Desiree bore him seven kids, and all of them were chiral, just as their father once was. Apparently, MeX had not eliminated the genetic causes of his chirality.

Authors: Thank you, Ralph. It's a beautiful and romantic story about passionate love, where the unfortunate chiral goat, in order to marry, had to utilize the Sadowsky effect. Our readers probably know about it too. But we'll remind them anyway. The Sadowsky effect is the emergence of a mechanical rotating torque on a disk irradiated with circularly polarized light (1898). In other words, if a stationary black light disk capable of rotating frictionlessly around a vertical axis is irradiated with circularly polarized light, after some time, the disk will begin to slowly rotate in the direction of the rotation of the plane of polarization of the falling light. Simply put, the angular momentum carried by the circularly polarized light is transferred to the disk absorbing this radiation.

Diogenes: I'm somewhat surprised by where a simple goat got such profound knowledge in physics.

Ralph: But he wasn't just any goat, he was a chiral goat. Now, if you, dear Diogenes, were a chiral cat, what would you feel?

Diogenes: No, no. Thank you. Do you want my tail to rotate only in one direction, e.g., only clockwise? No, and no again, I tell you in response.

Ralph: I would like to note that if the tail rotates only in one direction, there's nothing terrible about it. For example, the flagellum of a human spermatozoon rotates only in one direction. Therefore, when spermatozoa bind to egg cells, they begin to rotate the egg cells counterclockwise. Although the reason for the asymmetry of this rotation is not clarified, it can be asserted that we are dealing with the chirality of the spiral beating of spermatozoa.

Authors: Thank you for the interesting addition. All that remains is to conclude that chirality pursues us everywhere: from DNA molecules and spermatozoa to the very horns. And indeed, it is one of the distinctive and mysterious properties of life. For example, to this day, scientists cannot explain why the snail *Fruticicola lantzi* with a shell in the form of a right spiral (typical) is more viable than a snail with a shell twisted in the opposite direction (Fig. 21.1).

Figure 21.1 Chirality selection for life. From left to right, at the bottom: homochiral molecules of the left-handed alanine, DNA right-handed helix, left–right asymmetry of cell division, right (typical) form of the snail Fruticicola lantzi which is more viable than the inverse form, left–right asymmetry of the human body, and left–right inversion in the human brain under influence of a magnetic field. Importantly, all amino acids are present in all proteins only in the left configuration.

Chapter 22

Flounder: An Artist and Pharmacist

Meet me, a young shark. My name is Tiburon. In the ocean, I have no enemies except orcas. And I am beautiful, strong, and fearsome with two rows of teeth. I am the queen of these waters. I eat whom I want and when I want.

But something strange happened to me recently. Let's start from the beginning.

My grandmother used to tell me that they often had flounder for breakfast in the old days. She said it was very tasty but a strange fish with unusual abilities. For instance, it's strange because it's born looking like a regular fish, with eyes symmetrically located on either side of its body and the same coloration on both sides. But as it matures, its eyes and some vital organs shift to one side. Eventually, adult flounders "lie down on their side" and continue their life in a horizontal plane. The fish swims with one side up and the other always facing down. The coloration of the flounder also becomes asymmetrical: its right side becomes white or yellowish, while the left side, which becomes its upper part, can spontaneously color itself in any pattern. Speaking in the language of physics, the flounder undergoes spontaneous symmetry breaking of the body. The entire body of the fish tilts to the left side: here are the eyes, only the left side of the body is colored, the fins are shifted, even the internal organs are arranged asymmetrically. This is a true left-handed fish! Right-

Shadowless Squids: Stories of Physics in Nature
Vitalii Zablotskii and Tatyana Polyakova
Copyright © 2025 Jenny Stanford Publishing Pte. Ltd.
ISBN 978-981-5129-43-4 (Hardcover), 978-1-003-57062-2 (eBook)
www.jennystanford.com

sided individuals are also found, but very rarely. The flat, wide, and short body is almost always positioned flat. The flounder also swims flat, moving its fins. It lives on the bottom, under the thickness of water, hence it's flattened. Interestingly, flounders were even found at the bottom of the world's deepest Mariana Trench. Flounders move slowly along the bottom surface using oscillatory movements of their fins. But if they sense danger, they literally "turn on their edge" and swiftly swim away in that position.

I knew all this, but can you really escape from a shark? After all, I can swim at speeds of up to 40 km/h. So, one cloudless morning, I decided to have a hearty breakfast. I spotted a slow-moving, fat, and tasty flounder weighing about 30 kg (the largest flounders can grow up to 4 m and weigh up to 330 kg) on the bottom and began to think about the best way to attack it. The flounder's body was covered with fine scales, but it had sharp bony spines on the sides of its body. I made my decision and started to gain speed, rapidly approaching it. But suddenly, the sand on the bottom stirred up, and the huge flounder disappeared as if it dissolved in the bottom water.

"What kind of strangeness is this," I thought and began to carefully examine the area of the bottom where the flounder had just been. The sand was just sand with small pebbles and algae on the surface. No signs of the flounder. Perhaps I imagined that the flounder was here. I swam further away and started looking for another flounder. I was really looking forward to breakfast. But suddenly, the sand behind me came to life and stirred. Out of it emerged the same flounder, darting away toward the coral reef. The flounder swam very fast, like a normal fish, i.e., standing on edge. I chased after it, but it was already hiding in the thick corals. "No, you won't hide from me here," I thought. Indeed, a 30 kg fish is easy to spot among small coral bushes.

But you won't believe it, it disappeared again! I swam over these coral bushes dozens of times, but I always saw the same scene: corals, mollusks, algae, and no flounder. I was getting tired of this game of hide and seek.

Finally, I realized that the flounder could literally blend in with its surroundings. It can quickly change the color of its upper side of the body depending on the color and pattern of the bottom; it reproduces it so perfectly that it becomes invisible. This magical ability is called mimicry. It's camouflage of the highest level.

I wonder, how does it manage that? The fish "scans or reads" the pattern and color of the bottom with its eyes. From the eyes, this scan as an electrical signal goes to the color control center—the spinal cord. After appropriate software processing, the image of this pattern is transmitted via nerve fibers to special cells on the body surface—chromatophores. This happens roughly the same way as transferring a photo from a computer to a color printer. Chromatophores change the color of the upper side of the flounder to camouflage in just 5 min. But to most accurately reproduce all the nuances of its surroundings, the flounder needs several hours.

"So, that's it, it hid somewhere here. Keep looking," an inner voice told me, and I started to carefully examine each coral. One of them seemed suspicious, and I poked it with my nose. But suddenly, a small jet of some slime squirted right into my nose. Instantly, I felt difficulty breathing and a slowdown in my circulation. It was poison that squirted from a gland located near the flounder's fins.

At this point, I was no longer interested in breakfast. I had to get away from this disgusting fish. Breathing heavily, I swam away from that place and never hunted flounders again. Perhaps my grandmother was talking about a different kind of flounder, or I didn't listen to her attentively. After all, the flounder family is quite substantial: it includes more than 500 species.

That's where my adventure with the flounder ended, and I switched to a healthier diet. I went on a diet, even two, since one isn't enough to fill me up.

Reader: Does this mean that a repellent for sharks can be created based on flounder venom?

Authors: It seems so. Scientific research has confirmed that the mucus from flounders, when released into the water, repels sharks for 10 h. Biochemists have determined that the mucus contains a toxin called pardaxin, which causes difficulty in breathing and circulation for sharks. This toxin kills sea stars, urchins, and other enemies of the flounder. To observe pardaxin's effect on sharks, scientists conducted an unusual experiment. In a laboratory pool, they threw a small flounder directly into the open mouth of a shark. And the shark remained with its mouth open, as the toxin caused a spasm in its muscles. And the flounder calmly swam away. One of the scientists even tested the effect of pardaxin on herself by taking a drop of the toxin orally. Fortunately, nothing terrible happened; she only felt a slight burning sensation on her tongue. Therefore, this substance is safe for humans and can be developed into a shark repellent. With just a concentration of 0.8 mg/l, sharks turn away in flight.

Ralph: I have an addition regarding the strange asymmetry of left and right in the flounder's body. To varying degrees, the asymmetry of left and right exists in most organisms. For example, the shells of the snail *Fruticicola lantzi* are most often twisted to the right. It's strange that snails with right-handed spirals are more viable than those with left-handed ones. And no one knows why.

Diogenes: Yes. The asymmetry of left and right manifests in various aspects of the life activities of living organisms and humans, up to the realm of psychology. For example, the visual perception of Raphael's "Sistine Madonna" significantly changes when reflected in a mirror.

Figure 22.1   Does anyone see a flounder here?

Reader: For me, the most amazing thing is how the flounder photographs and accurately reproduces the pattern of the sea bottom on its back. Fishermen say that if you put a flounder on a newspaper, within 5 min its coloration will closely resemble the newspaper text. It's as if the flounder has a camera on its stomach and a high-quality printer installed on its back (on the upper side) that quickly and efficiently prints photos even in seawater.

Authors: The mechanism of operation of chromatophores—special cells in the flounder's skin that can change its color—is more akin to the mechanism of action of a liquid crystal display of a computer monitor. The flounder has several types of chromatophores, each responsible for a specific color. When the flounder wants to change its color or camouflage itself to blend into the surrounding environment, it uses the "photograph" of the sea bottom and its nervous system to control the chromatophores. Signals from nerve endings reach the chromatophores and lead to changes in the size and distribution of pigment granules inside the cells. This mechanism allows the flounder to be a master of camouflage and adapt to various environments for better survival.

Reader: It would be great if someone could develop a method of applying tattoos based on the chromatophores of flounders, transplanted into human skin! Then you could change the pattern and color of the tattoo several times a day.

Authors: We will not comment on this. Just note that, in addition to the chromatophores, you would need to transplant the flounder's nervous system into humans.

Dear reader, if you decide to do this, my advice to you is to take the nervous system from the *Mimic Octopus*, which lives on the border between the Pacific and Indian Oceans. Its nervous system is more advanced because it simultaneously controls eight arms, and the octopus, like the flounder, has cells—chromatophores. Unlike many animals that change the color of their bodies to escape predators or attract partners, the *Mimic Octopus* is an absolute champion of mimicry because it can change its shape to look like a completely different creature. Its ability to mimic other organisms is astonishing. It can imitate and appear as lionfish, jellyfish, and sea snakes. Interestingly, among all its mimicries, its most preferred camouflage is that of a flounder. Such an advanced level of defensive mechanism indicates the high intelligence of this octopus and an advanced thought process.

Chapter 23

Picnic at a Beer Bar

Have you ever seen crows drinking beer? No, this didn't happen in the tale of the crow that lives 300 years by an ancient castle and greets colorful tourists. Everything took place near a regular beer bar in one of the northern countries.

You probably know that crows are quite intelligent birds. From time to time, they gather in one place, e.g., outside the city, to discuss pressing issues of domestic and foreign policy. They can sit there for hours and discuss something, but no one has yet learned the crow language, so it remains a mystery what they are actually talking about. Maybe about physics? We think that crows also know physics quite well. That's what our story will be about.

Once in a small town near a beer bar, I saw several drunk crows. They walked strangely, clumsily shifting from one foot to the other and even tried to fly. But flying didn't work out well for them: they flapped their wings loudly and couldn't gain altitude.

"Drunk crows!?"—I thought. Could such a thing even be real? But it wasn't a dream. And I decided to investigate this unusual phenomenon. At first, I thought that there might be a biological laboratory nearby where experiments on birds were conducted. But besides the beer bar and residential houses, I found nothing noteworthy in the vicinity. And then I saw an unusual sight.

Shadowless Squids: Stories of Physics in Nature
Vitalii Zablotskii and Tatyana Polyakova
Copyright © 2025 Jenny Stanford Publishing Pte. Ltd.
ISBN 978-981-5129-43-4 (Hardcover), 978-1-003-57062-2 (eBook)
www.jennystanford.com

Not far from the beer bar, crows were drinking beer. Let me tell you what it looked like.

It was a winter Friday evening, and the bar was crowded, probably stuffy inside. Therefore, people came out with beer in half-liter plastic cups and drank beer right in front of the bar. The used plastic cups were thrown into a nearby container, but it was already overflowing. Many empty and partially filled cups were next to the container, right on the snow. And it was these partially beer-filled cups that became lawful prey for the crows.

Figure 23.1   We are enjoying beer ice.

"But where is the physics here?"—you might ask.

And I will answer you with a question. How can crows drink from a cup? Don't know? But they do. It's very simple and, one might say, ingenious. It was winter outside, and the crows sat and waited for the beer to undergo a phase transition from liquid to solid. Simply put, they waited for the beer to freeze. But they didn't just wait for the phase transition in the beer, since solid ice is very hard to peck at. They were even smarter than you think. They controlled the beer's crystallization process, trying to catch its initial stage. For this, the crows tipped the cup with beer remains into the snow and waited for the liquid beer to turn into yellow fluffy snow—an intermediate phase between liquid and ice. It was this yellow snow that they pecked at, not drank.

Those who had already pecked enough beer tried to fly. But it was not to be. Alcohol made its presence felt. Some jumped up and flapped their wings clumsily, and those who managed to fly to the nearest tree dozed off on it.

Crows do not drink beer in natural conditions, as alcohol can be harmful to them. However, in some cases, crows may accidentally consume beer if it's available and they are near places where people consume alcoholic beverages. Crows like to explore their environment and try various objects, including open bottles or cans of beer.

Reader: Amazing! Not just crows, but real researchers. Did they snack on cheese with their beer?

Authors: No. Crows eat cheese in other places, e.g., in fairy tales.

Reader: And I read that crows know mathematics. For example, to fly from point A to point B, located at a certain distance from each other and at different heights, crows always choose a flight trajectory that is the shortest of all possible. Is that true?

Authors: Why not? But we will not prove it here. Let's leave this question to mathematicians. Moreover, crows that have had their fill of beer are unlikely to be able to calculate the optimal flight trajectory.

Reader: Can I ask another physics question? At what temperature does beer freeze? At the same temperature as water, i.e., at 0°C?

Authors: No. Beer usually starts to freeze at about minus 2 °C. However, the exact freezing temperature can vary depending on the alcohol content and other additives in the beer. But the main point is not at what temperature this particular beer freezes. The general rule states that the freezing temperature of a solution is lower than the freezing temperature of a pure solvent. The change in the freezing temperature of a solution can be calculated using Raoult's law: the change in the freezing temperature of the solution is proportional to the molarity of the solution. But you will learn this in a thermodynamics course at the university. So, at temperatures higher than minus 2°C, crows do not drink beer.

Reader: So, crows have already attended university and know Raoult's law?

Authors: Perhaps they know it from birth on an intuitive level.

Chapter 24

How a Fly Walks on the Ceiling

Here we talk about how a fly can walk on the ceiling and window glass. If the gecko, a descendant of an ancient lizard, uses modern nanotechnologies to walk on the ceiling, then the fly gets by with knowledge of school physics. Let's try to explain this.

It's all about surface tension and the Laplace formula.

Diogenes: I got it all. Enough already.

Ralph: Probably, many of our readers have also understood everything by now. But I would listen. Repetition is the mother of learning.

Authors: Let's start with the fact that the free surface of a liquid behaves much like a thin elastic film. Surely, you've seen how some insects, like water striders (*Gerridae*), run on water without sinking. If you look closely at the water's surface directly under the legs of this insect, you will see that it has bent, i.e., formed a small depression. The surface of the water behaves like a stretched flat film and does not allow external forces to deform it. This is a manifestation of surface tension forces.

To understand why surface tension forces arise, let's first consider a molecule inside the liquid, and then the same molecule at the surface of the liquid, bordering, e.g., with air. Inside the liquid, the molecule interacts with other molecules evenly surrounding it from

Shadowless Squids: Stories of Physics in Nature
Vitalii Zablotskii and Tatyana Polyakova
Copyright © 2025 Jenny Stanford Publishing Pte. Ltd.
ISBN 978-981-5129-43-4 (Hardcover), 978-1-003-57062-2 (eBook)
www.jennystanford.com

all sides. Thus, the resultant force acting on this molecule equals zero. Remember, force is a vector quantity. And if you add vectors of the same length, emanating from one point and evenly distributed in space, you get zero. If you mentally move this molecule to another point inside the liquid, the situation does not change. Thus, the state of the molecule inside the liquid remains unchanged, and therefore it has the same energy everywhere, as long as it is inside the liquid. But for molecules near the liquid's surface, the situation is entirely different. Here, the molecule is surrounded by neighbors only halfway. Below are water molecules, and above are air molecules, with which it hardly interacts since the air density is significantly less than water's density. It turns out that on the surface, a water molecule interacts with half the number of neighbors compared to the same molecule but inside the liquid. Therefore, the resultant force acting on our molecule from the neighbors is directed inside the liquid.

Reader: Sorry, but you promised a story about flies, and you're talking all about molecules.

Authors: We'll gradually get to the flies.

So, where were we? Ah, yes. On the fact that inside and on the liquid's surface, molecules are in unequal conditions. Since the potential energy of attraction forces is always negative, the molecules on the surface have more energy than molecules inside the liquid. Therefore, near the surface, a force acts on the molecules, returning them to the liquid if they want to move perpendicular to the surface. And now the most important part. If you want to stretch the surface or bend it, it means you will be increasing the liquid's surface area. On a molecular level, increasing the area is equivalent to increasing the number of surface molecules. Indeed, by increasing the surface area, you move some molecules from the volume to the surface, thereby increasing the system's energy. Like any physical system, the liquid strives to reduce its energy, meaning to reduce the area of its free surface. As a result, forces arise, called surface tension forces, which strive not to allow the stretching of the liquid's surface. For example, a blown soap bubble takes the shape of a sphere. Do you know why? Because the surface of a sphere has one remarkable property, which is easily proved mathematically. Among all bodies of equal volume, a sphere has the minimum surface area.

Diogenes: Are you saying that if we lived in a world with a different geometry, where, e.g., a cube had the minimum surface area of all bodies of equal volume, soap bubbles would take the shape of a cube. Right?

Ralph: Obviously, yes. But you, Diogenes, with your philosophical questions, are distracting readers from the main narrative.

Authors: Or maybe on the contrary. Diogenes, with his questions, expands the reader's horizons.

Figure 24.1 Walking on the ceiling is such a delightful experience.

Let's continue. Surface tension is a force that arises on the surface of a liquid when its area is increased. Each liquid is characterized by a surface tension coefficient. The surface tension coefficient σ is numerically equal to the force acting per unit length of the "liquid–gas" interface boundary and striving to reduce the liquid's surface. This force is directed tangentially to the interface boundary and perpendicular to the perimeter. The measurement unit for σ is N/m. Now we're just a few steps away from getting to the flies. It's important to note that surface tension forces change the pressure under a curved surface—the meniscus. Moreover, if the meniscus is concave, the pressure under it is less than atmospheric. And if the meniscus is convex, the pressure under it is greater than atmospheric, as the surface tension forces are directed inward of the surface. The difference in pressure above and below a curved surface is determined by the Laplace formula. This excess pressure is called Laplace pressure.

To verify the existence of Laplace pressure, perform a simple experiment. Take two small glass plates. Drop some water between them and then press the plates together. Now try to separate them. It turns out, they are attracted to each other with quite a significant force, equal to the Laplace pressure (ΔP) multiplied by the wetted surface area of the plate (S): $F = \Delta PS$. Water wets the glass surface and forms two concave menisci, as shown in this drawing. Thus, according to the Laplace formula, the pressure between the plates is less than atmospheric by the amount of ΔP, depending on the surface tension coefficient of water and the main curvature radii of the meniscus: R_1 and R_2.

Diogenes: If any of the readers doesn't know what the radius of curvature of a surface is, ask the crocodile.

Ralph: What jokes you have, Diogenes. You'll scare all the readers away. You could just say normally: read our story about the frostbitten crocodile.

Authors: Since the pressure between the plates is reduced, atmospheric pressure presses the plates together. Note, according to the Laplace formula, this force is greater, the higher the liquid's surface tension coefficient. And the fly knows this, as well as the formula $F = \Delta PS$, and successfully uses it. The surfaces of the fly's feet are wetted by a special liquid—fly grease, which has a quite high surface tension coefficient, σ. It's precisely the high surface tension coefficient of fly grease and the Laplace formula that play a key role in the fly's ability to walk on the ceiling or glass.

Ralph: But that's understandable, the area of the fly's feet is small. It means that to increase the adhesion force with the ceiling, one needs to increase σ.

Authors: Scientists have calculated that the total force, $F = \Delta PS$, with which the fly's feet press against the surface of the ceiling or wall is about 2.4 mN. Since the fly's own weight is about 0.7 mN, this force is quite sufficient to keep the fly on the ceiling.

Formulas used by a fly:

Laplace pressure (general case)	Pressure difference between two slabs	Pressure difference in spherical shapes such as bubbles or droplets
$\Delta P = \sigma\left(\dfrac{1}{R_1} + \dfrac{1}{R_2}\right)$	$\Delta P = \dfrac{\sigma}{R}$	$\Delta P = \dfrac{2\sigma}{R}$
R_1 and R_2 are the principal radii of curvature, σ is the surface tension	$R_1 = R$ but $R_2 \to \infty$	$R_1 = R_2 = R$

Chapter 25

Crime at the Hotel

Let's talk about how animals use nanotechnology.

Diogenes: "He who owns nanotechnologies controls the world."

Ralph: Fortunately, cats haven't mastered nanotechnologies yet.

Diogenes: Unfortunately, not yet. But you, dogs, are far from nanotechnologies as well.

Ralph: Not as far as you'd like to think. I'll tell you about an incident that happened to me.

We were vacationing on one of the paradise islands in Southeast Asia. The vegetation, the sea, the people, and the wildlife— everything was different from our country, giving the feeling that we were either on another planet or, traveling through time, ended up in our planet's past. Everything was perfect, but at the hotel where we stayed, the police showed up every morning for some reason. The officers inspected one room or another and talked to the guests. As a lover of detective stories, I immediately realized that it was about thefts from the rooms. Later, I learned that almost every night, jewelry mysteriously disappeared from the rooms: gold rings, diamond brooches, earrings, and even necklaces. These were all things that women took off at night and placed on a bedside table, usually located next to the bed. By morning, these items vanished without a trace. The police assured that the hotel was constantly

Shadowless Squids: Stories of Physics in Nature
Vitalii Zablotskii and Tatyana Polyakova
Copyright © 2025 Jenny Stanford Publishing Pte. Ltd.
ISBN 978-981-5129-43-4 (Hardcover), 978-1-003-57062-2 (eBook)
www.jennystanford.com

under the surveillance of external and internal video cameras, and therefore the criminal had no chance to enter a room and steal anything.

I pondered over these crimes, trying to find even the smallest clue. The main thing was that there were no traces or witnesses, but the crime existed. Police dogs sniffed the rooms and the corridor, but all in vain. Even an expert in paranormal phenomena was called in. But he couldn't say anything sensible.

During a walk, my attention was drawn to a small dragon skillfully running up the trunk of a smooth tree. It was a gecko—a special breed of lizard that has remained almost unchanged from prehistoric times. It was the first time I saw such a large lizard and knew nothing about them. Meanwhile, the gecko climbed down the tree and ran up the vertical wall separating the beach from the road. And at that moment, an idea flashed through my mind like lightning. What if it was the gecko stealing the jewels in our hotel? But why would it need them? Absolutely not. But, on the other hand, it is well known that magpies steal shiny objects and store them in their nests. But they do it out of interest, not out of a desire for wealth. And here we're talking about items worth hundreds of thousands of dollars. Why would a gecko need such treasures? These and other questions swirled in my mind. And I was about to discard this thought when I spotted the gecko again.

I took a photo of this dragon and looked up information about it on the Internet. Here's what I found: size 10–30 cm, weight 50 g, color variable, harmless, non-venomous, feeds on small rodents, difficult to tame compared to all other types of lizards, can make sounds. The sounds they make resemble the sound of... GECK. Gecko's feet are covered with numerous microscopic hairs that cling to the support surface through van der Waals forces, helping the lizard to move across the ceiling, glass, and other surfaces. A gecko weighing 50 g can hold a load of up to 2 kg on its feet.

Oops... This is exactly what I needed. The Internet is indeed useful. After that, the picture of the crime became clear to me. The gecko enters through an open window into the room and steals jewelry from the table, moving exclusively along walls and ceilings. Now it's clear why the cameras don't see it, and the dogs can't pick up its scent sniffing the floors in the room and corridors. The only question left was, why does it do this? And what if someone trained it and taught it to steal jewelry in such an original way for its owner?

In the evening, I shared my thoughts with the police inspector. And that very evening, the police installed several video cameras right on the tables, where the jewelry was placed. In the morning, as expected, some of the jewels disappeared. But when the inspector reviewed the footage from the cameras, there was no doubt left. In the middle of the night, a fairly large gecko came along the wall, took the necklace, and also left along the wall through the window.

A few days later, the criminal was found and arrested. It turned out to be the caretaker of the local zoo, who had been caring for geckos for many years. And, as it turned out, he was a good trainer.

Figure 25.1　Perfect thief.

Diogenes: Can someone explain to me how this criminal manages to walk on the ceiling? I, for one, can't do it.

Ralph: That's because you don't possess nanotechnologies.

Authors: Allow us to explain this from a physics point of view. The first thing to remember is that in our world, there is matter in a condensed state (i.e., solid and liquid, but not gaseous). This is due to intermolecular interaction forces. These forces manifest themselves in such a way that molecules, which somehow find themselves close to each other, are attracted to each other, but if the distance between them is made smaller than the diameter of one molecule, attraction turns into repulsion.

This can be understood if you imagine that the molecules are connected by small springs. When we increase the distance between neighboring molecules, the spring stretches and a force arises, striving to return the molecules to their previous places. And when we bring the molecules closer, the spring shortens, and now the force, directed in the opposite direction, again strives to return them to their original position.

Usually, in a solid body, molecules are packed so that the average distance between them is about several diameters of a molecule. But at slightly greater distances, molecules begin to noticeably attract each other.

Diogenes: Well, according to this theory, I should be able to walk on the wall. After all, the molecules of my paws are attracted to the molecules of the wall!

Authors: Hold on. Let's first imagine that your paw is touching the wall. Mentally look at the paw touching the wall. See, here the surfaces of the paw and the wall are shown magnified. Notice how uneven these surfaces are. When they come into contact, only a small portion of the molecules of both surfaces come into close contact (meaning the molecules are brought within a few diameters of each other).

Diogenes: Ah, I see, only those at the peaks of the micro-hills are in contact?

Authors: Yes, exactly. But unlike human feet or cat paws, gecko feet are covered with numerous microscopic hairs that adhere to the surface via van der Waals forces. Therefore, all the valleys between the micro-hills are filled with the gecko's nanohairs, and the contact area of its feet's nanohairs with the surface is very large. At the same time, each pair of attracting molecules creates a force on the order of $f = 10^{-12}$ N.

Ralph: Is that a large force?

Authors: No, it's a very small force. For example, to hold a 1 kg weight, a person needs to apply a force of 10 N, i.e., a force 1000 billion times greater than the force of attraction between two molecules. Now you see that to make the molecular attraction force between the paws and the wall surface sufficiently large, it's necessary to significantly increase the number of molecules that are in close contact, i.e., at very small distances from each other. But at the micro-level, all real surfaces are very rough, and for this reason, most of the molecules on the contacting surfaces cannot come close together. Thus, the total force of attraction between the surfaces will be negligibly small.

Ralph: I understand, to increase the attraction force, one must polish the surfaces very carefully. And then they will stick together.

Authors: You're right, carefully polished surfaces indeed attract each other with a sufficiently large force. But in such a case, it would be very difficult to separate them. Therefore, this method is not suitable for geckos.

However, the structure of the external surfaces of their feet, covered with nanohairs, allows them to significantly increase the number of molecules in contact, providing a sufficiently strong adhesion force between their feet and the surface of the wall or ceiling.

Each gecko foot is covered with the thinnest hairs, only a few nanometers thick (a millionth of a millimeter). When such a foot touches a surface, each nanohair finds its part of the wall surface and creates a tight contact. Knowing the number of molecules in contact with the surface, scientists calculated the total force with which a gecko adheres to the wall or ceiling. It turns out that this force is about 40 N, i.e., it allows holding a load of about 4 kg! Therefore, a gecko can freely rest on the ceiling or walk on it. So, it can be said that the gecko possesses one of the secrets of nanotechnology, which is very trendy in the scientific world now. We hope we've explained everything clearly?

Diogenes: Clear, it's clear, but not everything. Let's say you've convinced me that the force of attraction of a gecko's feet to the ceiling or wall can be so large that it allows holding up to 4 kg. But how can it run? After all, now it somehow has to detach its feet from the ceiling? Is it that strong?

Authors: Good question. But our answer is very simple. Remember how we peel off tape stuck to the wall. To peel it off, we pull on the corner and detach it gradually, step by step, by a small area of the surface. And if we try to peel off the entire piece of stuck tape at once, i.e., detach its surface from the wall so that the tape remains parallel to the wall all the time? Of course, we won't succeed because the force required would be very large. The gecko does the same to detach its feet from the walls; it first tilts them and detaches its nanohairs in separate, comparatively small groups, small bundles. Understand?

Diogenes: Yes. I understand that I won't be able to walk on walls and ceilings until I grow nanohairs on my paws.

Ralph: I don't understand you, dear Diogenes. First, you wanted to grow huge wings like a condor to hunt bats. And now you want to grow nano-hairs to walk on the ceiling. Why?

Diogenes: I want freedom of movement in 3D space. And you all keep proving to me that walking on the ceiling is impossible. But here comes a fly and shows a master class in walking on glass and the ceiling. Are you going to tell me now that the fly possesses nanotechnologies?

Authors: The fly arrived just in time, at the end of our story. But the fly's ability to walk on the ceiling is related to another physical phenomenon, which is called—the surface tension of liquids. But we've already told you about that.

Chapter 26

The Largest Organism on Earth

Do not trust your eyes.

—Kozma Prutkov

Lyrical

A green forest and a glade, filled with white mushrooms. This idealistic picture paints itself in the imagination of anyone who has ever picked wild mushrooms. Nature, coolness, and pleasure. And no physics! After all, these are just mushrooms—some of the simplest and most ancient plants (organisms) on our green planet. Mushrooms can grow anywhere: from desert dunes to the frozen tundra. It's best to collect mushrooms in the forests after the rain. It's a pleasure and a simple algorithm consisting of three steps: walk slowly, see a mushroom, put it in the basket.

But, are they as simple as they seem at first, second, and even the nth glance?

Facts

What if we immediately say that a mushroom is the largest organism on Earth? Don't believe it? Then who is the largest?

Shadowless Squids: Stories of Physics in Nature
Vitalii Zablotskii and Tatyana Polyakova
Copyright © 2025 Jenny Stanford Publishing Pte. Ltd.
ISBN 978-981-5129-43-4 (Hardcover), 978-1-003-57062-2 (eBook)
www.jennystanford.com

Reader: The largest organism on our planet is the blue whale. Its mass reaches 150 tons, and its length is 33 m. And the maximum mass of a mushroom, probably, is 1 kg.

Authors: Let's listen to what scientists have to say about this.

Scientists refuted the notion that the blue whale is the largest living organism on the planet, discovering in Michigan (USA) a honey fungus, weighing 440 tons and occupying an area of 37 ha. The mass of this fungus is equal to the mass of three whales! Scientists estimate its age at 2.5 thousand years.

You might ask, how can a mushroom be so big?

Answer: What we call mushrooms are just the fruiting bodies of these sophisticated and mysterious organisms. The main part of the mushroom is a mass of branched underground threads with a thickness of 0.5–10 μm, called mycelium.

Thus, the majority of this gigantic honey fungus is hidden underground and, as we said, occupies 37 ha = $3.7 \cdot 10^5$ m², which approximately corresponds to a circle with a diameter of 1200 m! The mycelium of this honey fungus weighs about 400 tons. Depending on the type, mushroom mycelium threads can feed on substances in the soil, decaying plants, or wood. For large honey fungi, the mycelium threads are quite thick and can grow meter by meter in search of wood for their nutrition. While other mushrooms prefer already rotting wood, honey fungi infect living trees, often killing them over several decades, and then continue to eat them after the trees have already died. If you are collecting honey fungi, a sure sign that the mushroom is under your feet is a grove of trees dying above it.

It was experimentally established that mycelium can gather and systematize information about the surrounding environment, understands its location in space, and transmits this information to its offspring—parts of the mycelium that were separated from the mother mycelium. To prove this, scientists studied the movements of the mycelium of the simplest fungus (*Physarum polycephalum*) in search of food in a labyrinth. It turned out that even such a simple organism can find the shortest path between two points in a labyrinth. Could it possibly know Fermat's principle? The French mathematician Pierre Fermat formulated the fundamental principle of geometric optics in 1660, now called Fermat's principle. According to this principle, out of all possible paths between two points, light chooses the one that takes the least time to travel. Considering that the mycelium threads spread at a constant speed, it can be concluded that mycelium, like light, chooses the path that requires the least time.

This means that at the cellular level, the fungus is capable of making mathematical calculations and can demonstrate primitive intelligence. This is very astonishing! On this basis, we can assert that mushrooms are the smartest living organisms in the world. It sounds like an exaggeration, but anything is possible.

Another stroke, speaking of the intelligence of fungi. External digestion—they came up with that well. In this process, enzymes are first released into the surrounding environment containing food substances, which outside the organism (not inside) break down polymers into easily absorbable monomers, which are then absorbed into the cytoplasm. The release of enzymes and absorption of nutrients are effective only in moist soil. That's why mushrooms grow intensively after the rain. It rained, now you can eat. Of course, we're talking about mushrooms, not people.

Physics: Electric mushrooms

Mushrooms belong to the group of microorganisms capable of generating electricity as a result of their metabolic processes. Each mushroom has a certain electric voltage between the tip of its stem and its cap, which can be measured by inserting electrodes into the mushroom.

During their growth and differentiation, fungal cells generate both steady and varying electric currents (action potentials). Additionally, they demonstrate two effects of the electric field: (1) galvanotropism—the bending of growing roots or plant shoots under the influence of a constant electric current passing through the surrounding environment, and (2) galvanotaxis—directed movement determined by the magnitude and direction of the electric current. Moreover, natural constant electric currents with a density of up to $0.6 \, \mu A/cm^2$ flow through fungi. These currents may arise due to the clustering of ion channels and ion pumps in certain cell areas of the mycelium. It appears that electric currents and action potentials are associated with the spatial control of nutrient absorption and intramycelial exchange.

The generation of electric current and response to it is an important and common aspect of fungal physiology. For instance, it is known that in phytopathogenic fungi, motile zoospores (simplest single-celled organisms that move in a liquid medium by beating one or several flagella) exhibit galvanotaxis in weak electric fields, comparable to fields generated by plant roots. This effect allows fungi to direct their zoospores and create their concentrations in places where there are natural electric fields generated by plant roots. This means that fungi can use the endogenous electric currents of plants to detect their potential victims.

Just like that! Some fungi recognize their victim in the same way sharks do—by their electric signals.

Many generations of Japanese mushroom farmers have noticed an increase in mushroom yield in places struck by lightning. Experiments conducted with mushrooms and artificial lightning have confirmed this phenomenon. However, a clear explanation for this effect is yet to be found. It can only be presumed that possessing primitive intelligence, fungi react to lightning as a significant threat to survival and therefore accelerate their growth.

Thus, during their growth, fungi exhibit electrical activity in the form of changes in their electric potential. Also, fungi respond to external stimuli, such as chemicals, by altering their internal electric currents and voltages. It can be said that the organism of fungi lives under the control of electricity. But do fungi feel pain? Perhaps. Scientists, studying the electrical signals of the fungus *Pleurotus ostreatus* (common oyster mushroom), discovered a phenomenon that can be interpreted as fungal anesthesia. The electrical activity of *Pleurotus ostreatus* is characterized by slow (hours) irregular waves of change in the resting cell potential and fast bursts (minutes) of the action potential. Treating the mycelium-substrated vapor with chloroform led to a several-fold reduction in the amplitude of the baseline potential waves and an increase in their duration. Chloroform vapors also caused either a complete cessation of electrical activity or a significant reduction in the frequency of electrical spikes. Removing chloroform vapors from the growth containers gradually restored the mycelium's electrical activity. Isn't this akin to anesthesia in fungi?

And to leave no doubt in anyone's mind that fungi are very intelligent and simply filled with electricity, we will present the results of recent research here. It turns out that fungi can communicate with each other using words and even sentences. Their vocabulary includes up to 50 words—electrical impulses of a special shape, spreading through the mycelium network. Mathematical analysis of electrical currents in the mycelium revealed that each species of fungi has its own recurring electrical signals (like words), which form sentences.

Figure 26.1   The kingdom of mushrooms millions of years ago.

Living batteries: Mushrooms generating electricity

Biologists from the Stevens Institute of Technology in the USA have created mushrooms, based on ordinary champignons, capable of conducting electricity. Scientists already knew that cyanobacteria *Anabaena*, residing at the ocean's bottom, could generate current both in light and darkness. However, studies were complicated by the fact that *Anabaena* are sensitive to unfavorable conditions and die quickly.

Scientists managed to solve this problem. They combined cyanobacteria with special nanoparticles, placing them on the surface of mushrooms between strands of graphene, which serve as accumulators and conductors of electric current. In other words, a thin graphene net with cyanobacteria was draped over champignons. Why champignons, specifically? It turns out that for *Anabaena*, the champignon provides a complete set of necessary nutrients. Moreover, the moisture level and temperature in it are ideal for these microorganisms.

Under the influence of sunlight or lamp light, the champignon with cyanobacteria generated an electric current of about 65 nanoamperes. Of course, this is very little to power any device, but, if several mushrooms are combined together, they can indeed provide electricity for a small lamp.

Reader: This is a revolution in "green technologies"! Imagine if many cyanobacteria-covered champignons are placed under the hood of a car, the electricity they produce would be sufficient to rotate the shaft of the electric motor of such an absolutely clean car. Just water the mushrooms under the hood and go wherever you want without any emissions.

Authors: Yes, you've proposed a fantastically ecological car. Perhaps in the future, biotechnologies will be able to make significant progress in such an "engineering symbiosis" of bacteria and fungi.

Scientists designed mushroom computers

We have already mentioned that a data transmission mechanism using electrical signals was discovered within mycelium! And scientists developing computers based on new principles of computation and data transmission immediately took advantage of this discovery. It's not surprising, given that mycelium is a mathematical graph. Recall, a graph as a mathematical object is a collection of two sets—the set of objects themselves, called the set of vertices, and the set of their pairwise connections, called the set of edges. What is needed to design a computer? Correct, some physical signals, for example in binary code, representing "zero" and "one."

If we take an "electrically excited" mushroom (with a burst of oscillations)—as a one, and a non-excited one—as a zero, then it's possible to try to implement some logical operators of Boolean algebra on the mycelium. Scientists did this, albeit only in computer simulation so far. They presented the mycelium with four mushrooms as a graph, where the vertices, besides the mushrooms themselves, were also the intersections of the mycelium. There were a total of 2×10^4 nodes in the simulation and were denser the closer they were to the front edge of the mycelium. Scientists managed to implement the logical "not" on these four mushrooms by applying stimulation or its absence to them and then, after some time, measuring the voltage on these same mushrooms. Clearly, these are only the first steps of a new era of computers. And their performance is still far from the performance of semiconductor processors. But the idea works.

Reader: Wow! It turns out, there's a "computer" growing under every fir tree in the forest.

Authors: Yes, you can go into the forest and collect "computers." Just be careful, many of them are poisonous and deadly dangerous!

Reader: In the end, mushrooms generate internal electricity, react to external electric fields and currents, and even use electrical signals to hunt for prey. And also, they are real forest computers.

Authors: Dear reader, why is your cat so silent today, not asking any questions?

Reader: My cat listened to the beginning and declared that he's not planning to gather mushrooms. That's what my philosopher said and then went for a walk. And as he left, he added, "Now, if mushrooms grew directly in sour cream..., that'd be a different story." Meanwhile, my dog listened to everything attentively and is literally dragging me to the forest for mushrooms. My dog even started getting ready. I had to promise him that we'd go to the forest this weekend. So, my dog started to learn the scents of the tastiest mushrooms.

Authors: Oh, it seems we've found a true connoisseur of our stories.

Chapter 27

The Oldest Electric Microcheetah

> *You have to run as fast as you can just to stay in place,*
> *and to get somewhere, you have to run at least*
> *twice as fast!*
> —Lewis Carroll, *Alice in Wonderland*

Archaea are the ancient organisms that seem to have lost their way among the trees and bushes of evolution. Indeed, archaea (from the Latin Archaea, derived from the Ancient Greek ἀρχαῖος, meaning "ancient, original, old") are one of the forms of life alongside bacteria and eukaryotes. It's fascinating that archaea, being single-celled microorganisms without a nucleus or any membrane-bound organelles, have their own independent evolutionary history and are characterized by many biochemical features that distinguish them from other life forms. Some publications have noted that remnants of archaeal lipids were found in the Isua greenstone belt in western Greenland, home to the Earth's oldest sedimentary rocks, formed 3.8 billion years ago. Thus, archaea could be the oldest living beings on Earth. However, it remains a mystery whether archaea are the ancestors of eukaryotes. The biological peculiarities of archaea could be discussed endlessly.

Shadowless Squids: Stories of Physics in Nature
Vitalii Zablotskii and Tatyana Polyakova
Copyright © 2025 Jenny Stanford Publishing Pte. Ltd.
ISBN 978-981-5129-43-4 (Hardcover), 978-1-003-57062-2 (eBook)
www.jennystanford.com

Let's view archaea through the eyes of a physics engineer. Can you imagine a cheetah running at 3000 km/h? Probably not. But in the world of archaea, something similar exists. Most archaea are chemoautotrophs, utilizing a significantly broader range of energy sources than eukaryotes, from common organic compounds like sugars to ammonia, metal ions, and even hydrogen. It's as if archaea feed on what seems inedible, such as metals and hydrogen. Oops..., hydrogen is an environmentally friendly fuel. When it burns, it produces clean water instead of soot and harmful gases. This is why scientists and engineers worldwide are actively working on developing a hydrogen engine. Imagine driving a car that emits only water from its exhaust pipe. While humans are still working on the hydrogen engine, archaea have been using it for hundreds of millions and even billions of years!

Another type of archaea, the haloarchaea, use sunlight as an energy source, which is also very modern and eco-friendly. Archaea can live in very harsh conditions, such as hot springs (geysers), black smokers, and saline lakes, as well as under high pressure—up to 700 atmospheres. Their membranes, composed of glycerol-ether lipids, maintain a liquid crystalline state across a wide temperature range (0–100°C), crucial for their biological functioning. This helps archaea survive at high temperatures, as well as in highly acidic and alkaline environments. They can also live in soil, oceans, swamps, and even in the human gut. Fortunately, archaea are very friendly: none of the known archaea is a parasite (except for nanoarchaeota, which parasitize other archaea) or a pathogenic organism. Planktonic archaea might be the most numerous group of living organisms currently on Earth.

Archaeal cells range from 0.1 to 15 μm in size and can have various shapes: spherical, rod-like, spiral, or disk-like. Archaea move using flagella, which, like in bacteria, are driven by a rotational mechanism located at the base of the flagellum. This mechanism operates on electricity—transmembrane proton gradient. However, the flagellum (archellum) of archaea is structurally different from that of bacteria. Using high-resolution cryo-electron microscopy, scientists have closely examined the flagellum of the tiny archaeon *Methanocaldococcus villosus*—one of the planet's fastest organisms. It turns out that this archaeon's flagellum is made up of thousands of copies of two alternating proteins. This flagellum, like an electric motor, can rotate at very high speeds. Rotation of the archellum clockwise moves the archaeon forward, while rotation counterclockwise moves it backward. The flagellum's rapid rotation allows this archaeon to move at a speed of 0.5 mm/s. Considering that an archaeon is only 1 μm long, it can cover a distance equal to 500 lengths of its body per second. For comparison, a 1.4 m long cheetah running at 110 km/h (about 30 m/s) covers a distance equal to about 20 lengths of its body per second. If *Methanocaldococcus villosus* were the size of a cheetah, it would swim at a supersonic speed of 2750 km/h, making it one of the planet's fastest organisms. Remember, the speed of sound in the air is 1191 km/h.

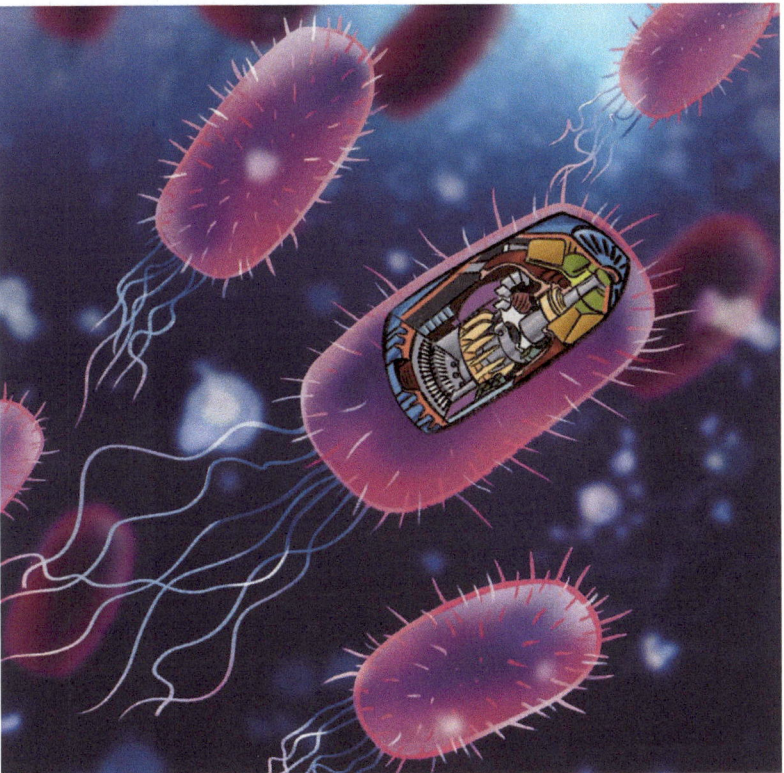

Figure 27.1   The oldest electric motor within.

Reader: This comparison with a cheetah is interesting, but one of your stories mentioned that a cheetah cannot run at high speed for long. If a cheetah does not catch its prey within 1 or 2 min, it stops the chase. How about archaea?

Authors: You're right, a cheetah can only reach maximum speed for a short time because its body temperature increases significantly (up to 40–42°C) under intense exertion, which can end very badly for it, as its blood might clot, leading to instant death. This temperature increase results from the intensive work of all its muscles. Moreover, as you already know from previous stories, animals, like humans, use a chemical reactor (stomach) as an energy source, where food combustion (reaction with oxygen) occurs slowly, accompanied by heat release. Unlike the cheetah, archaea have an electric engine and a non-thermal energy source. They cannot overheat because their temperature always equals the temperature of the surrounding liquid (in thermodynamic equilibrium with the environment), and they do not need to chase anything. Except perhaps a sugar molecule wandering in the liquid as a Brownian particle. But if an archaeon is on a diet, it can be satisfied with ammonia or hydrogen.

Reader: Are you suggesting that if a cheetah were powered by electricity and had the same kind of portable electric motor as archaea, it wouldn't overheat as quickly?

Authors: Electric motors are much more efficient than internal combustion engines, hence they generate less heat. But if your cheetah were powered by electricity like an electric car, it wouldn't need to chase prey. Well, maybe just for fun.

Reader: Your comparison of archaea's speed to that of a cheetah is interesting but not entirely accurate. Archaea move in liquid, while a cheetah runs in air. And we know that the resistance to movement in water is about 800 times greater than in air. Could you compare the speed of archaea to, e.g., a dolphin?

Authors: Nothing could be simpler. A dolphin, with a body length of 2.3–3.7 m, can reach speeds of up to 50 km/h ($\approx$14 m/s). This means a dolphin covers approximately 5 body lengths/s, while an archaeon covers 500 body lengths. Thus, if *Methanocaldococcus villosus* were the size of a dolphin, it would swim at a speed of 5000 km/h. Or to put it another way: if a dolphin were the size of an archaeon (1 µm), it could swim at a maximum speed of only 0.005 mm/s.

Reader: Ah, then it wouldn't survive the high temperatures, immense pressures, and high-speed water currents that prevailed on Earth during the emergence of life. The behavior of an organism in such extreme conditions must follow the rule: "relocate and seek."

Chapter 28

Frostbitten Crocodile with Flat Eyes

Diogenes: Authors, judging by the title, you want to present us with a thriller. Then I'm leaving, as I don't like thrillers. I'd rather go to sleep.

Ralph: Diogenes, don't go to sleep, or you might dream of a frostbitten crocodile with flat eyes.

Diogenes: Yes. You're right, that would be horrible. Honestly, I'm not fond of crocodiles.

Ralph: I also don't like thrillers. And besides, I don't have any crocodile friends.

Authors: Dear Dr. Diogenes and dear Ralph, please stay and listen to the story until the end. As always, we await your sharp questions and critical remarks. This is not a thriller, but a short story about the biophysics and physiology of a crocodile.

Let's start with the fact that the brain of an adult crocodile weighs only 8 or 9 g and is the size of a walnut.

Diogenes: To say the least, that's not much.

Ralph: But that doesn't mean crocodiles are stupid.

Diogenes: Yes, I read somewhere that a bee makes decisions faster than a human.

Authors: Indeed, it has long been noticed that crocodiles are quite smart and even use tools to lure prey, such as sticks to attract birds. The human brain weighs 14001500 g, and an elephant's brain

Shadowless Squids: Stories of Physics in Nature
Vitalii Zablotskii and Tatyana Polyakova
Copyright © 2025 Jenny Stanford Publishing Pte. Ltd.
ISBN 978-981-5129-43-4 (Hardcover), 978-1-003-57062-2 (eBook)
www.jennystanford.com

weighs 5 kg. But weight is not the main characteristic of the brain. From the perspective of physics and computing, the brain is a highly efficient portable computer that consumes only 20 W of energy. That's about the same power as a mobile phone charger. But for a modern computer to perform as many computational operations per second as the human brain, it would need to have the power comparable to a nuclear reactor!

Diogenes: Are you suggesting that the crocodile has a very small brain to save energy?

Authors: Most likely not. Nature simply gave it a brain according to the tasks it needs to solve in its life.

Ralph: And its tasks are simple: eat someone bigger.

Authors: And precisely for this, evolution gave the crocodile flat eyes. Or more accurately, eyes with almost flat surfaces of the pupil.

Diogenes: Is that to scare the prey?

Authors: No. It's about physics, specifically geometrical optics. If you have the patience, we will explain this in detail. So, the crocodile's lifestyle is hunting. And predominantly hunting for prey that swims in water. Although crocodiles can run fast on land, up to 20 km/h, water is their native element.

So, imagine yourself in the crocodile's place. You're swimming in the river and see some animal crossing the river. That's food, says your 8 g brain. You speed up to 40 km/h and quickly catch up to the prey. It's time to attack. But in the heat of the chase, you haven't had a chance to properly look at your prey. What if it's some kind of monster? You need to take a good look at it from all sides. And here's the main question.

For a successful attack, a crocodile needs to see its prey equally clearly both under and above water. Nature gave crocodiles eyes that are elevated above their snout and half-submerged in water when hunting. Moreover, crocodiles have eyes with horizontal oval pupils, placed side by side in their sockets, allowing them a wide field of vision and improved binocular visual perspective. This enables them to accurately determine the distance to the target and effectively hunt both on land and in water. But do crocodiles see equally well in water and air?

You've probably noticed that when you dive with open eyes, all objects underwater look blurry, i.e., out of focus. This happens because the human eye, specifically the lens, which acts as a converging lens, is designed by nature to provide a clear image on the retina. In other words, the image should be in focus. Or as photographers say, the camera lens (like the eye lens) should be focused on the subject for the photo to be clear. You might not notice this because modern cameras have autofocus, but professional photographers know this well. Now imagine your camera lens is half in the water, and you want to take a picture both underwater and above water. Autofocus won't help you here. And you surely don't know what to do.

But the crocodile knows. It knows the thin lens formula, and that's why it has flat eyes. Look at this formula at the end of the story. It includes n, the refractive index of the lens material (camera lens or eye lens); n_0, the refractive index of the medium; and R_1 and R_2, the radii of curvature of the lens's front and back surfaces, respectively. Remember, for air $n = 1$ and for water $n = 1.5$.

Ralph: Sorry to interrupt. But I'm not sure all our readers know what curvature and the radius of curvature are.

Authors: From a mathematical point of view, curvature is a measure that characterizes the deviation of a surface from being flat. If a certain part of a curved surface is represented as part of a sphere, then the radius of this sphere is called the radius of curvature of that surface. Note that according to this definition, a plane has an infinitely large radius of curvature.

Diogenes: That should be clear to everyone. After all, we all live on Earth—a big sphere with a radius of 6400 km—and it seems flat to us.

Authors: As always, you're right, Diogenes. But let's get back to our crocodile and its cunning eyes. So, it knows the formula for a converging lens and wants to have an equally sharp image of its prey on its retina, both under water and above water. But it is known that the refractive indices of water ($n_0 = 1.5$) and air ($n_0 = 1$) differ significantly. This means there will be a big difference in the focal lengths of its eye in water and in air (Table 28.1).

Ralph: Are you saying the focal length of a camera lens in water and in air would be different?

Authors: Yes, that's correct for both a camera lens and the eye's lens. But the crocodile's goal isn't to photograph its prey but to catch it. If, e.g., the image of the prey obtained in water falls on the retina and is sharp, then the image obtained by rays coming from the air won't be on the retina and will be significantly blurred. To correct these images, something from the lens focal length formula must be changed. The crocodile naturally cannot change the refractive indices of air, water, and the eye's lens. What does nature, in the form of a crocodile, do in this case? It only has to change the radii of curvature of the crocodile's eye surfaces so that the difference in distances where images are formed in water and air is minimal.

Table 28.1 Formulas known to the crocodile

The focal length, F, of a lens	The thin lens equation
$$\frac{1}{F} = \left(\frac{n}{n_0} - 1 \right) \left(\frac{1}{R_1} - \frac{1}{R_2} \right)$$	$$\frac{1}{a} + \frac{1}{b} = \frac{1}{F}$$
where n is the index of refraction of the lens material, n_0 is the index of refraction of the medium, and R_1 and R_2 are the radii of curvature of the two surfaces.	shows the relationship between the image distance (b), object distance (a), and the focal length (F) of the lens.

For this, the crocodile applied (somewhere in the process of evolution) another thin lens formula that you learned in school, and mathematically proved that the images in water and air would be equally sharp (or nearly so) if the radii of curvature of its pupil were sufficiently large.

Diogenes: Ah, so that's why it has flat eyes.

Ralph: Almost flat. A large radius of curvature means a slightly curved surface.

Authors: Exactly right. A surface with low curvature appears almost flat and has a large radius of curvature. We leave it to advanced readers to prove on their own that a converging lens with large radii of curvature for the front and back surfaces has only a slight difference in focal lengths in water and air. Hint: to prove this, play with the two lens formulas presented at the end of the story.

To summarize, the shape of crocodile eyes is the result of their evolution and adaptation to their habitat. One of the main reasons why crocodiles have flatter eyes compared to other reptiles is to maintain good vision both underwater and on land.

Diogenes: Somehow, with flat eyes and a 9 g crocodile brain, it all started to make sense. Flat eyes for better simultaneous vision in water and air, and a small brain is enough for solving simple life tasks like catch and eat.

Ralph: Plus, such a brain is resistant to temperature drops and even freezing. A crocodile can freeze so that it becomes encased in ice in a frozen lake. Then, after thawing, it comes back to life, looks good, and seems perfectly healthy.

Authors: Alligators can survive at temperatures down to 5 °C. But if it gets colder, their body enters hibernation: all metabolic processes slow down, food isn't digested, and all energy is spent on maintaining minimal body function at ambient temperature, e.g., for breathing. Thus, alligators frozen in ice are a normal occurrence. This is how reptiles survive low temperatures. They go into hibernation, and to avoid suffocating underwater, they only stick their snouts out from under the ice.

Diogenes: What a delightful spectacle. You go for a walk on a frosty day to the lake, and right out of the ice here and there, frozen crocodile snouts are sticking out. No, it still seems more like a bad dream or a thriller.

Authors: Scientists assert that crocodiles "sense" when water begins to freeze and react by timely pushing their nostrils to the surface. This process is called brumation—a state in which an animal's body temperature and metabolic rate drop significantly. Brown bears also have the ability to hibernate at low temperatures. But as a warm-blooded animal, the bear needs nutrients, even in small amounts, during hibernation. Therefore, a bear sucks its paw during sleep, through which some nutrients previously stored are ingested. Thanks to this, a bear can sleep for several months. A frozen crocodile during hibernation doesn't eat but only breathes. Therefore, alligators cannot stay frozen for long: a week at most. If the weather doesn't warm up within a week, they can die. But if a thaw quickly follows the frost, crocodiles gradually thaw and revive.

Ralph: I can't imagine what a frostbitten crocodile looks like. Does it feel okay? Maybe it caught a cold while stuck in the ice?

Authors: Clearly, long freezing is not beneficial for alligators. Scientists haven't yet asked a frostbitten crocodile how it feels. Note that crocodiles don't die of old age because they don't biologically age. They die from hunger or diseases.

Figure 28.1 The frozen crocodile is waiting for the thaw.

And here's something interesting. Scientists studying American alligators found that the temperature at which a crocodile's eggs are laid determines the sex of the future crocodile. In American alligators, females hatch at 30 degrees Celsius, and males at 33 degrees Celsius. This is regulated by the protein TRPV4, present in the gonads (sex glands) of alligator embryos. Depending on the surrounding temperature, it allows calcium ions into cells, activating genes that lead to male development.

Diogenes: So what does this mean? If the sex of future crocodiles is determined by the ambient temperature, will we only have male crocodiles in case of global warming?

Ralph: Obviously, that's what will happen. But that would be the last generation of crocodiles on Earth since males don't lay eggs.

Diogenes: Allow me to disagree with you. The future could be different. You don't know nature well enough.

Authors: Let's leave this debate and return to crocodiles. I have a question for you.

What do you do if a crocodile is chasing you?

Note that on land, it runs quite fast, up to 20 km/h = 5.6 m/s, faster than the average person. And in water, it can reach speeds comparable to a motorboat and catch almost any living being.

Diogenes: I'm not planning on meeting a crocodile.

Ralph: Of course, you won't meet them. It's hard to do that lying all day on the couch. And I understand that, whether you meet a crocodile in water or on land, your chances of escaping it are close to zero.

Authors: But there are two secrets, based on the physics and physiology of the crocodile, that could help you escape.

Listen and remember them.

If a crocodile is chasing you on land, run away from it in zigzags. Why? Because the crocodile has a large mass (and therefore, significant inertia) and cannot quickly change its direction of movement. In this case, making sharp zigzags, you are likely to escape.

And if a crocodile chases you in water? Here the advantage is entirely on its side. Your only option is to dive underwater every time the crocodile opens its mouth to grab you. Remember, a crocodile cannot open its mouth and catch prey underwater. But be careful, it might hit the prey with its powerful tail.

Ralph: I saw on YouTube how one gazelle, escaping from a chasing crocodile, dived underwater a few meters from the shore. By doing so, it won this deadly race since the crocodile cannot catch prey underwater.

Diogenes: Interesting, why can't a crocodile catch prey underwater?

Authors: A crocodile cannot catch prey underwater due to certain physiological limitations of its anatomy. First, most crocodile species have teeth on the upper jaw used for capturing prey at the water's surface. Second, closing the nostrils and a special valve at the back of the throat help the crocodile hold its breath and not drown underwater, but these limit its ability to open its mouth wide and catch prey underwater.

Diogenes: Very useful information for tourists. When I'm in Africa, I'll definitely look into a crocodile's eyes. Are they really flat? I heard that if you flip a crocodile upside down, whether in water or on land, it just stays lying there because it probably doesn't understand what happened to it. It seems to live in a 2D space.

Ralph: You're probably going to Africa without getting off the couch, right? And when I'm in America, I'll definitely visit a nature park in North Carolina to see the lake with crocodiles frozen in ice.

Chapter 29

Extreme Bear: Questions Without Answers

I would rather have questions that can't be answered
than answers that can't be questioned.

—R. Feynman

A creature known as the tardigrade, also called "water bear," was first described in 1773 by the German pastor J. A. Goeze as Kleiner Wasserbär (from German—"little water bear").

Of course, it's not actually a bear. It earned such a "nickname" because it moves extremely slowly, about 2–3 mm/min, and its walk visually resembles the leisurely gait of a bear swaying from side to side. Below, we present some facts characterizing this "bear"—the tardigrade.

Simple biology

The body of tardigrades ranges in size from 0.1–1.5 mm, is semi-transparent, consists of four segments and a head. For instance, a tardigrade from the class *Eutardigrada* has a body length of 200 μm and a mass of 23 μg.

According to archaeological data, tardigrades have lived on Earth for more than 500 million years.

Shadowless Squids: Stories of Physics in Nature
Vitalii Zablotskii and Tatyana Polyakova
Copyright © 2025 Jenny Stanford Publishing Pte. Ltd.
ISBN 978-981-5129-43-4 (Hardcover), 978-1-003-57062-2 (eBook)
www.jennystanford.com

They are found everywhere, from the Himalayas (up to 6000 m) to ocean depths (below 4000 m). Tardigrades have been found in hot springs, under ice, and at the ocean floor. They also occur in freshwater bodies, but most inhabit moss and lichen cushions on the ground, trees, rocks, and stone walls.

Tardigrades feed on the fluids of algae and other plants they inhabit. Some species eat small animals: rotifers, nematodes.

They have digestive, excretory, nervous, and reproductive systems; however, they lack respiratory and circulatory systems. The role of blood is performed by fluid filling the body cavity.

The active life of tardigrades usually ranges from 3–4 months to two years among different species. But in a state of anabiosis, they can survive for decades.

Tardigrades primarily survive through a process called anhydrobiosis, or drying out. When drying, they retract their limbs, decrease in volume, and take on a barrel shape. The surface is covered with a waxy shell that prevents evaporation. During anhydrobiosis, their metabolism falls to 0.01%, and their water content can reach 1% of the normal. They dry out and stop showing signs of life, their metabolic processes slow down by several million times. In this state, tardigrades can withstand extreme temperatures, dehydration, radiation, and pressure, waiting for more suitable conditions for active life.

But the most interesting fact is that tardigrades can survive in extreme physical conditions. Tardigrades are the most resilient creatures on Earth.

Life in the extremal physical conditions

Temperature. They withstand a 30 year stay at a temperature of –20 °C; 20 months in liquid oxygen at –193 °C; 420 h at a temperature of 0.01 K (almost absolute zero); heating to 60–65 °C for 10 h and to 100 °C for an hour.

Radiation. Tardigrades can survive doses of radiation equal to 10,000 Gray (Gy). The lethal dose for the bacterium *Escherichia coli* is 50 Gy, for mice 7 Gy, and for humans 4 Gy. Tardigrades can repair DNA segments damaged by radiation. They contain a special protein Dsup (DNA damage suppressor protein), which protects their DNA from damage, including radiation damage.

Pressure. "Sleeping" tardigrades were placed in a high-pressure chamber, gradually increasing it to 600 MPa (about 6000 atmospheres), and after this monstrous pressure, they came back to life. Note that the pressure in the deepest part of the ocean—the Mariana Trench—is only 110 MPa!

Outer space. In an experiment, tardigrades spent 10 days in outer space. After 10 days spent in outer space in a state of anabiosis, almost all organisms were desiccated, but aboard the space station, all tardigrades returned to normal and were able to produce normal offspring.

Figure 29.1 I can travel across the Universe without a spacesuit.

The main questions to which the authors do not know the answers

Why do tardigrades need such super-abilities to survive in extreme conditions?

If tardigrades acquired their super-abilities through the process of evolution, one wonders where they encountered such extreme conditions (immense pressures, high doses of radiation, ultra-low temperatures, and outer space) to adapt to them?

What genetic programs and molecular mechanisms allow them to recover after surviving cataclysms?

Diogenes: The question "why?" is more of a philosophical question. There can be no definitive answer to it. Science does not answer questions like "why?" "for what purpose?" Let's better discuss the question—"how?" How did they acquire these abilities to survive practically in any conditions?

Ralph: You can't adapt to something you've never experienced.

Diogenes: I also don't understand how this super creature adapted to live in harsh cosmic conditions. After all, to acquire such qualities through evolution, one needs to live for a sufficiently long time in such conditions. And, probably, more than one generation of tardigrades would have to pass through, e.g., open space. This means that the ancestors of tardigrades traveled long distances through foreign galaxies, from planet to planet, before arriving on Earth.

Reader: Maybe they arrived from distant worlds on a comet, which, according to one hypothesis, brought water to our planet? So, they directly arrived in the water.

Ralph: Okay, let's assume they arrived on a comet with water. But being in water on a comet and being in open space are different things. Moreover, this comet must have passed somewhere near a powerful source of radiation, e.g., near a black hole.

Diogenes: If your comet had been caught in the gravitational field of a black hole, it would have fallen into the black hole along with your tardigrades.

Ralph: Well, they definitely wouldn't have survived in a black hole.

Reader: Could horizontal gene transfer have played a role? It's believed that tardigrades might be capable of horizontal gene transfer—the transfer of genetic information between individuals without reproduction. This could facilitate rapid adaptation to new conditions and the acquisition of new abilities.

Authors: Friends, there's room for imagination here. Many questions will remain unanswered here. We really do not know how tardigrades acquired their unique abilities to survive in extreme conditions. Further research and genetic analysis may help clarify these mechanisms and provide a more complete understanding of the processes underlying their remarkable adaptability. We're confident that our readers will find answers to all their questions when they grow up and become researchers working in the field of biophysics.

Chapter 30

Street Quiz: Elephant in Questions and Answers

We decided to hold a science contest among random passersby on the street. We prepared a few simple questions about animals and took to the streets. We asked questions and didn't expect scientific answers because it felt more like a children's game of riddles rather than a biophysics exam. So, let's move on to the questions.

Question 1

Mister in the gray hat, could you please explain to us why elephants have big ears?

- Perhaps it's so they can hear better.
- And do you think that a cat, with its small ears, hears much worse than an elephant?
- No, I don't think so. My cat hears very well. Sometimes she hears the conversations of mice under the floor of our house. And it really annoys her.
- I see. Thank you.

Miss, one moment, please. Could you explain to us why elephants have big ears?

- What do you think? Imagine: a huge elephant with tiny ears on its head. It would be disproportionate and unattractive.

Shadowless Squids: Stories of Physics in Nature
Vitalii Zablotskii and Tatyana Polyakova
Copyright © 2025 Jenny Stanford Publishing Pte. Ltd.
ISBN 978-981-5129-43-4 (Hardcover), 978-1-003-57062-2 (eBook)
www.jennystanford.com

Young lady, excuse me. We have a question for you. Why do elephants have big ears?

- It's to flap their ears to ward off insects.
- Logical. Thank you.

Question 2

Young man, please answer one question. Who exerts greater pressure on the surface of the sidewalk, a 4 ton elephant or a 60 kg girl in stiletto heels?

- Obviously, the elephant.
- Did you study physics in school? And how did you guess that the elephant exerts greater pressure than the girl?
- Of course, I studied, but didn't understand everything. And it's very simple to guess. For example, if a girl steps on my foot, as a polite person, I'll keep silent, but if an elephant steps on me, it will hurt a lot, and I'll scream.
- Thank you. Interesting answer.

Sir, please answer one question. Who exerts greater pressure on the surface of the sidewalk, a 4 ton elephant or a 60 kg girl in stiletto heels?

- Can I see the girl?
- No. This is a question about physics, not physiology.
- Then I don't know. Physics is a dark science.
- Ah, I see, you're under the impression of dark matter, about which physicists are now so hotly debating.

Question 3

Young lady, please, one question. Do you know why an elephant has whiskers on its trunk?

- What are you talking about! I've never seen that. But if they grow, then they must be necessary.
- Thank you. Your answer is logical.

Miss, please, can you tell us why an elephant needs whiskers on its trunk?

- That's a tough question. Probably for the same reasons men have whiskers under their noses. After all, an elephant's trunk is an elongated nose fused with the upper lip.
- Thank you. We will think about why men need whiskers. Young man. Do you know why an elephant has whiskers on its trunk?
- Probably for the same reasons that rats have whiskers.
- And why does a rat need whiskers?
- Rat whiskers are their main tactile organ. Rats have poor vision, so whiskers help them feel objects and the environment, determine their texture, shape, and temperature. Whiskers allow rats to navigate in the dark or in places with limited visibility.
- Very interesting. Yes, elephants have relatively poor vision compared to other animals. But they don't crawl through burrows where it's dark and need to feel potential restrictions with their whiskers. Moreover, a rat's whiskers are only slightly shorter than its body size. But an elephant's whiskers are tiny compared to its size. So, your explanation doesn't work.
- Well, then I can't add anything more.
- Thank you.

Question 4

Sir, may I ask you one question?

- Please.
- Why do you think elephants always have one tusk shorter than the other?
- Because living nature is asymmetric with respect to left and right. In biology, this is called chiral purity. It manifests from the simplest biological molecules to the level of the organism and even our consciousness. In this case, elephants are no exception.
- Thank you for such a scientific answer. Clearly, you're a biologist.
- You guessed it. My profession is molecular biology.

Young man. Why do elephants always have one tusk shorter than the other?

- Probably because elephants often fight among themselves and break each other's tusks in the process.
- And do you, by any chance, practice martial arts yourself?
- Yes, I do karate.
- I see. Thank you.

Question 5

Young man. A small question about physics. An elephant weighs 5 tons. Can an elephant swim?

- Probably, but not on its back.
- Excellent.

Young lady. Excuse me. Do you think elephants can swim?

- I've seen elephants dance in the circus. They probably graduated from a dance school.
- Yes, elephants are very musical. Sometimes elephants display amazing grace and synchronization in their movement, which might resemble dancing. But can they swim?
- I think not, they are very heavy. An elephant probably weighs as much as a tractor, and a tractor definitely doesn't swim. I've seen it myself.
- Thank you.

Question 6

Student, one question for you. What is the pulse rate of an elephant?

- An elephant is big, so it must have a fast pulse to pump blood up to a height of 5.5–6.5 m.
- What do you think, how many beats/min?
- Well, if a human has 60 beats/min, then an elephant, which is, let's say, 3 times taller, should have a pulse of 180 beats/min.
- Thank you. It's good that you didn't take part in creating elephants.
- Ah yes, that was a long time ago. I wasn't even born then. Sir, could you tell me what the pulse rate of an elephant is?

- I'm not a doctor, but I think it's high. An elephant does a lot of physical work. And when a person works, their pulse increases.
- Thank you.

Hello, miss. Are you a physics student?

- Yes. How did you know? It's written on your t-shirt, Einstein's famous formula: $E = mc^2$.
- Exactly. But now, this formula is written everywhere: on cars, in advertising brochures, and even on pizza.
- I have questions for you. What is the pulse rate of an elephant?
- What's the mass of an elephant's heart?
- Up to 20 kg.

Figure 30.1   Why does an elephant need a trunk?

- Well then. We can estimate the average heart rate of an elephant, i.e., its pulse. It is known that the natural frequency of oscillation is inversely proportional to the square root of the mass of the oscillator. Therefore, the greater the mass, the lower the natural frequency of oscillation. If the average human pulse is 60 beats/min, then the elephant's should be significantly lower. I think about 20–30 beats/min.

- Excellent. It's clear that you have studied the oscillations and waves topic well.

We asked six simple questions about elephants to people on the street, who were representatives of different professions and social groups. We're sure that our readers have already provided their original answers to these questions, and now it's time to give our versions based on modern scientific facts about the physiology of elephants.

Question 1. Why do elephants have large ears?

An adult elephant's ears can reach a width of 4 m. They use them as a natural fan to cool themselves down and to ward off flies. But this is not the main reason for the enormous size of their ears. The main thing is that the sounds made by elephants are in the infrasound range, which means their wavelength exceeds the range normally perceived by the human ear. Infrasound frequencies are below 20 Hz, corresponding to a wavelength of more than 17 m. They can also include some lower components with even greater wavelengths. From a physics perspective, a receiver of such long sound waves should have dimensions corresponding to the wavelength of the signal. That's why large detectors—large ears—are needed. It's impossible to register a long wave with a small detector. The same applies to wave sources. The transmitting antenna of electromagnetic waves or the working part of a sound wave generator usually has dimensions comparable to the wavelength they generate.

It's noted that elephants can make rumbling sounds that are typically in the range of 14 to 24 Hz. Human ears can't perceive these sounds because they are too low (long wavelength), and human ears are too small.

Question 2. Who exerts greater pressure on the surface of the sidewalk, a 4 ton elephant or a 60 kg girl in stiletto heels?

For a physicist, it's obvious that a 60 kg girl produces more pressure with her heels than a 4 ton elephant. As known, pressure is the force acting on a unit area, $P = F/S$. If you compare the total area of the four elephant's feet with the area of two stiletto heels, you'll see that their ratio is much greater than the ratio of the elephant's weight to the girl's weight. This means that the girl on stilettos exerts greater pressure than the average elephant.

Question 3. Why does an elephant have whiskers on its trunk?

The tips of elephant trunks are covered with stiff hairs. For the first time in 130 years, scientists have figured out why these enormous animals need small whiskers on their nose.

Elephant whiskers—thick and immobile hairs on the tips of their trunks—were first described back in 1890. But since then, researchers hadn't really considered why they are needed.

Scientists discovered that elephant whiskers significantly differ from rat whiskers: they are cylindrical (rats have conical ones, preventing them from getting stuck in surrounding objects), relatively thick, and sturdy.

Interestingly, during complex movements of the trunk, the whiskers remain immobile, and the authors of the study believe that whiskers are not for "feeling" but to prevent dropping and damaging objects held in the trunk. The presence of sensitive hairs allows precisely understanding how much force needs to be applied when gripping something with the trunk. This is important since elephants often hold plant food that can significantly change shape when compressed.

Question 4. Why do elephants always have one tusk shorter than the other?

Elephants, like people, can be left or right-handed. Therefore, they adapt to primarily work with either their right or left tusk. Thus, one tusk is shorter than the other because it wears out faster. By looking at how the tusks wear down, you can determine which tusk an elephant predominantly uses, i.e., whether it is left or right-handed.

Question 5. An elephant weighs 5 tons. Can an elephant swim?

Elephants swim excellently, but they cannot jump or run. They have two types of gait: walking and fast walking or running. Usually, elephants move at a speed of 2–6 km/h but can reach speeds of up to 35–40 km/h for a short time. According to Archimedes' law, a body submerged in a fluid experiences a buoyant force equal to the weight of the fluid displaced by the body. Thus, whether a body floats or not does not depend on its mass at all. Everything is determined by the average body density: the body's mass divided by its volume. If the average density of the body is less than the density of the fluid, the body floats on the surface of this fluid. The average body density of an elephant is less than the average density of water.

Question 6. What is the pulse rate of an elephant?

The physics student gave an absolutely correct answer to this question.

It only remains to add that an elephant's heart weighs about 20 kg and beats at a rate of 30 times/min. For comparison: the pulse rate of a rabbit is 200, and for a mouse, it's 500 beats/min. The heart of a blue whale beats only 9 times/min.

Chapter 31

Falling Cat, Sandwich, and Perpetual Motion Machine

If you have a cat at home, you surely know that cats love to knock various items off tables or cabinets. For example, you placed a vase with flowers on the edge of the table, and your beloved cat accidentally knocked it to the floor, pretending it all happened by chance. Or your favorite mug was on the kitchen table, and suddenly your cat, with a light swish of its tail, sends it crashing to the floor. Do you think all this happens by chance? Or is your cat just clumsy? But all cats do this.

And all this happens not by accident at all. You simply underestimate the abilities of cats. By knocking items off tables and cabinets, they study the laws of gravity. In particular, they are very interested in the laws of free fall. "Why?" you might ask.

Because cats, unlike humans, love to climb trees or walk on roofs, enjoying freedom. And from a tree or a roof, one can easily fall and get hurt if one does not know the laws of free fall in Earth's gravitational field. So, they have to learn these laws of mechanics in practice, as no one has yet written a physics textbook for cats or opened special schools for them.

Shadowless Squids: Stories of Physics in Nature
Vitalii Zablotskii and Tatyana Polyakova
Copyright © 2025 Jenny Stanford Publishing Pte. Ltd.
ISBN 978-981-5129-43-4 (Hardcover), 978-1-003-57062-2 (eBook)
www.jennystanford.com

Figure 31.1 Master class on falling from a height.

But how can knowledge of the laws of mechanics help a poor cat?

Firstly, cats are not poor, they are usually wealthy, and there are even very rich cats. Recently, the news reported, e.g., that one cat inherited 150 million British pounds. And this is not the only case.

Secondly, from a young age, cats not only quickly grasp practically all the laws of mechanics and even electricity but also learn to apply them in practice.

But let's consider bodies that are in free fall in Earth's gravitational field. Let such a body be a cat that accidentally fell off a roof. The cat's position at the start of the fall is shown in the illustration.

Reader: Oh, the cat is upside down! This is not the most fortunate start to a fall.

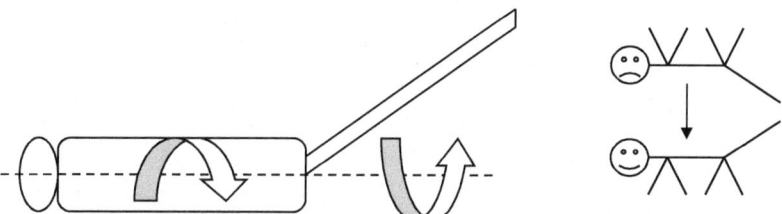

Figure 31.2   Schematic of falling cat.

Authors: Don't worry, this cat knows the laws of mechanics and uses them. And now we will tell you how it does so.

In the illustration, a cat that has fallen off a cabinet is schematically depicted in the "paws up" position. "Stay calm!" the cat tells itself. It's urgent to apply one of the fundamental laws of nature—the conservation of angular momentum: the angular momentum of a closed system of bodies (in which the moment of external forces relative to a stationary axis of rotation is zero) remains unchanged over time. Saying this, the cat starts to rotate its tail with an angular velocity ω_t. As a result, the cat's body begins to rotate in the opposite direction with an angular velocity ω_b. "Excellent, keep rotating," the cat thinks to itself and mentally notes down the law of conservation of angular momentum for the system cat + tail, which is in free fall: $J_b\omega_b - J_t\omega_t = 0$, where J_b and J_t are the moments of inertia of the cat's body and tail, respectively. In this equation, the minus sign before the second term indicates that the angular velocities of the cat's body and its tail have opposite directions. So, since the total angular momentum of the cat + tail system was zero before the start of the fall (note the zero on the right side of the equation), the sum of the angular momentum of the cat's body and tail remains unchanged (i.e., equal to zero) at any moment during free fall.

Now, the cat needs to know how long (t) it must rotate its tail in order to land on its paws, i.e., to turn its body exactly half a turn. For this, the cat needs to know the time of fall (t_0), which it can easily find from the formula: $h = gt_0^2/2$, where g is the acceleration of free fall and h is the height. It is clear that the inequality $t \le t_0$ must be met. For any cat, it's optimal to choose $t = t_0$ (think about why). The condition for landing on the paws is mathematically written as $\omega_b t_0 = \pi$ (during the time of fall, the cat's body will rotate by an angle of π, i.e., half a turn). The necessary angular velocity of tail rotation is calculated by the cat from the equation, knowing the formula for the moment of inertia of a cylinder $J = mR^2/2$, where m is the mass of the cat's body (or tail), and R is the radius of the cat's body (or the base radius of the cone described by the rotating tail).

Reader: And I know why cats choose the condition $t = t_0$. Obviously, if $t < t_0$, then they would have to rotate their tail very quickly, and from that, the tail might even come off.

Authors: Correct. To the cat, this is more than obvious.

So, our cat calculated all this correctly and successfully landed on its paws.

And now you have no doubts that cats learn, know, and love physics. We hope our readers will follow their example.

Reader: I understood everything. Thank you. But I have one question. You've proven that a falling cat always lands on its paws. Sounds like a law, right?

Authors: Well, if you want, you can call it the "cat's law."

Reader: Yes, let's call it the cat's law. But I know another law of nature. It's the sandwich law: a falling sandwich always lands butter side down. Heard of it?

Authors: Of course, we've heard and seen how it works in practice.

Reader: Now the question! What happens if you tie a sandwich, butter side up, to a cat's back? According to the "cat's law," such a cat should land on its paws, but according to the sandwich law, the sandwich should fall butter side down, i.e., the cat should fall on its back.

Authors: Let's not actually try this with a cat, okay? If we think about it, combining a cat with a buttered sandwich creates a funny situation. The cat wants to land on its feet, and the sandwich wants to land butter side down. If both try to happen at the same time, the cat and sandwich might just spin around in the air forever, because they can't decide which way to land. So, you've kind of invented a perpetual motion machine. Pretty cool idea, but let's keep it imaginary for the sake of the cat.

Once, one of the authors told his students about the problem of how a falling cat lands on its paws. In the next lecture, one of the students said the following. After your problem, I conducted several experiments with my cat. As a result, I no longer have a cat!

Was it an unfortunate landing? asked the lecturer.

No, the student replied. After those experiments, my cat just ran away from home.

So, don't experiment with your or any other cat. At the very least, you are obliged to ask the cat if it wishes to participate in such experiments. Although for those who have a cat, the cat's answer is obvious—NO.

Chapter 32

Underwater Electro-Hunting

It's fascinating to trace the evolution of hunting tools from ancient times to the present day. Let's start with humans. There was a time when humans were little different from animals and thus hunted with whatever was at hand: stones, sticks, and throwing darts. But gradually, evolution took its course, and as humans evolved (*or monkey upgraded to human?*), more sophisticated hunting tools appeared: spears, bows, and arrows. Initially, spears were just wooden, but then came spears with iron tips. Interestingly, the first iron spear tips were made from iron that had arrived on Earth in the form of meteorites. Later, gunpowder was invented, and guns appeared: first large and cumbersome, and later compact and automatic. And all this happened so quickly over just a few thousand years of human history. And what were animals doing during this time? Were they also improving their hunting tools? But looking at a tiger or a leopard, you wouldn't say so. What hunting tools do they have? Purely mechanical: teeth, claws, and swift paws. And they have had these for thousands of years. Perhaps their teeth and claws became stronger and sharper, but they remained qualitatively unchanged. Their weapon development stopped at the level of mechanics.

And snakes? Snakes invented venom, which is known to block the work of certain ion channels on the membranes of vital cells,

Shadowless Squids: Stories of Physics in Nature
Vitalii Zablotskii and Tatyana Polyakova
Copyright © 2025 Jenny Stanford Publishing Pte. Ltd.
ISBN 978-981-5129-43-4 (Hardcover), 978-1-003-57062-2 (eBook)
www.jennystanford.com

leading to the rapid death of the prey. This is already advanced chemistry and biology, not mechanics, you might say, and you would be right. And some snakes spit venom, not just anywhere, but right into the prey's eyes, so the venom hits the mucous membranes and acts faster. Here we see that knowledge of mechanics, biology, and chemistry is involved.

What other physics topics did we learn in school? Thermodynamics, electricity, and optics. We'll leave thermodynamics to the bees. We'll talk about optics and its role in hunting in a separate story. Here, we'll touch on the use of electricity as a hunting tool by animals.

From a physics perspective, hunting with electricity in the air atmosphere is impossible. No one has managed to do this yet, as it requires creating very high electric field strengths, such as those that occur before a storm. Then the air becomes a conductor of electric current, and electrical discharges—lightning—occur. At lower electric field strengths, air is a poor conductor of electric current, and thus, an electric discharge is impossible. But water, especially salty seawater, conducts electric current well. So let's look at the electrical methods of underwater hunting used by eels and electric rays.

The electric eel (*Electrophorus electricus*) uses an electric field to hunt small fish. It turns out that the eel can generate an electric field and cause electric discharges (or throw electric darts) around itself. It emits short (10–15 ms) pulses of a sufficiently strong electric field, causing the prey's muscles to contract spasmodically (tetanus). The eel can generate electric field pulses at a frequency of 400 Hz, creating a potential difference of up to 600 volts! By attacking and paralyzing its prey with such electric "shots," the eel calmly approaches it and then swallows it. "This is modern high-tech weaponry," you might say. Undoubtedly. Add to this the eel's ability to navigate by the Earth's magnetic field, and you have not just an eel, but a hunter equipped with the latest technology.

Reader: I have a question. Imagine, for a moment, that, e.g., a tiger, in the course of evolution, acquired the ability to throw electric lightning at its prey. Then it wouldn't have to chase antelopes. Just see an antelope, throw lightning at it, and lunch is ready. And this would be as much more eco-friendly as an electric car is more eco-friendly than a car with an internal combustion engine. Right?

Diogenes: How wisely nature acted, not giving tigers or leopards such abilities. Otherwise, they would have burned all the jungles with their hunting lightning. As for comparing the eco-friendliness of a regular car and an electric one, we'll talk about that later.

Ralph: I imagined this perfect creature, possessing modern navigation means and future weaponry. It's truly terrifying. But there's one question concerning human and animal evolution. As we know, primitive hunters used throwing darts and spears. Later, their arsenal included bows, crossbows, and then guns. How did the evolution of hunting weapons occur for eels? Did they skip all the previous stages of perfecting their hunting weapons and even outpace humans in this matter?

Authors: Yes, it's indeed a mystery. But you must remember that in water, any weapon like throwing darts, bows, and arrows is absolutely ineffective due to the high resistance of water. When a body moves in a fluid, it encounters quite a large resistance force, which is proportional to the speed of the body's movement. Try running into the sea, submerged in water up to your waist. It will be difficult. Perhaps that's why nature took a different path: the eel somehow learned to "throw" an electric field, which spreads in water at the speed of light and encounters no resistance from the water. One might say that such fish simply had no other choice but to tame electricity and develop electric weaponry.

Diogenes: Are you suggesting that if human evolution had occurred in water, people would have also learned to hunt by throwing electric fields at wild animals?

Authors: We do not know the answer to this question. We leave it to the readers. Thanks to its astonishing features and hunting strategy, the electric eel (*Electrophorus electricus*) is a formidable predator. When striking, it generates electric impulses with a voltage of hundreds of volts, lasting 1 ms, and a frequency of 500 Hz. A burst attack transfers electricity to the prey, stimulates its motor neurons, and consequently leads to tetanic muscle contractions and immobilization. The effect of the eel's attack is akin to being hit by a stun gun. The high frequency of discharges causes multiple muscle contractions and tetanus in the prey (or potential predator).

Pay attention to the simple physics behind the ability of freshwater and marine fish to generate electric impulses. In freshwater fish (fresh water has relatively low electric conductivity compared to seawater), most of the elementary batteries are connected in series, while in marine fish, they are connected in parallel. This is how the electrical organs of fish are adapted to life in freshwater and seawater. From your school physics course, you should remember that the EMF (electromotive force) of batteries adds up when they are connected in series, whereas in a parallel connection, the EMF of the entire battery remains equal to the EMF of each element, but the currents add up. Since freshwater has lower conductivity than saltwater, a higher potential difference (or higher electric field intensity) is needed to produce an electric discharge, which is achieved in freshwater fish by connecting the batteries in series. This is why the electric eel is long like a snake, and the maximum potential difference is achieved between its head and tail. Conversely, the electric ray is wide because it lives in seawater and can afford to connect its batteries in parallel within its body.

Let's repeat. With batteries connected in series (eel), their EMF adds up so that the total EMF equals the sum of the EMF of all the batteries in the series, but the current flowing through the battery equals the current of a single battery. On the contrary, with batteries connected in parallel (ray), the total current equals the sum of the currents of individual batteries, while the EMF of the battery remains equal to the EMF of a single battery. Knowing this rule, try to answer the question: which of the two fish, the river eel or the marine ray, would produce a higher voltage, and which a higher current?

Reader: Obviously, the ray should produce a higher voltage, and the eel a higher current.

Authors: Correct. The ray can output up to 20 V and 50 A, while the eel generates up to 600 V and, according to some data, up to 1200 V, but with a current up to 1 A. This example shows how simple laws of physics help us understand nature's design: how nature engineered these fish considering their different living conditions.

Reader: How are the electric batteries connected in an electric car? In series or parallel?

Authors: In an electric vehicle, batteries can be connected in parallel, in series, or even in a combined manner. It depends on the specific design and characteristics of the vehicle, as well as the desired battery configuration.

Interesting facts. Rays emit discharges in volleys, each consisting of 2–10 or more pulses, each lasting from 3 to 5 ms. The strongest electric discharges, with a power of up to 6 kW, have been found in the ray *Torpedo occidentalis*. For comparison, the power of a typical electric kettle is about 2 kW, with a current consumption of about 10 A. It has been calculated that 10,000 eels could generate enough energy to move an electric train for several minutes. However, after this, the train would have to stop for several days while the eels regenerated their electrical energy supply. This property can be used for catching eels (eel is an edible fish). But, as you see, catching it is dangerous. One method of catching relies on the fact that an eel, having discharged its battery, becomes harmless for a long time. Therefore, fishermen do this: they drive a herd of cows into the river and the eels attack them and deplete their electric charge. After driving the cows out of the river, the fishermen catch the eels. Interestingly, eels do not cause significant harm to the cows. Think about why.

Diogenes: Here's an absolutely clean and cheap source of electric power for you. No need to rack your brains over the problem of creating a thermonuclear reactor to charge millions of electric vehicles.

Reader: Hold on, you're not seriously suggesting we swap out Tesla's sleek batteries for a squirming mass of eels, are you? Imagine popping the hood to show off your engine and instead, there's just a bunch of slippery eels glaring back at you.

Ralph: Absolutely! Why not? After all, it's common knowledge that cats have a penchant for fish. Plus, think of the conversations at charging stations: "My eels pack's got more charge than yours!"

Reader: What a great idea, though. No need for wind turbines that kill birds, nor to build power plants that pollute nature. Instead, we should create eel farms everywhere. At gas stations, cars would swap depleted eels for fresh ones, and the tired and discharged eels would be sent back to the farm for rest and recharge (fattening).

Authors: Joking again. By the way, during an electric discharge, when current passes through the body of an eel or ray, some of its tissues noticeably heat up.

But let's get back to our fish. Electric fish use their superpowers not only as a weapon but also for electrolocation. Why is electrolocation necessary? Electric fish, like eels or electric catfish, often live in rather murky water. Moreover, the electric eel and electric catfish are typical nocturnal predators. Therefore, electric organs are used by fish in murky and dark waters for communication with a mating partner, navigation, prey detection, and defense. There are passive and active electrolocation.

Figure 32.1   Underwater electricity. Who is stronger, an electric ray or an eel?

All marine fish emit weak electric discharges that conduct well in the surrounding water. With passive electrolocation, the electric fish captures these signals and precisely determines the location of the prey, tracking all its movements to choose the most effective attack method.

Active electrolocation in electric fish is very similar to echolocation in bats. During active electrolocation, the fish sends out electric signals into the surrounding water. Any object within the electric field partially reflects these signals and distorts the signal, creating interference. These signal distortions are captured by the ray or eel using electroreceptors located on the surface of the skin. The parts of the fish's body with electroreceptors capturing the distorted electric field signal "project" an electric image, which, after being processed with a "special computer and corresponding program," provides the fish with comprehensive information about the object: distance to the object, its size, and shape.

Moreover, electric fish communicate with each other using an electric language, more precisely, through electric impulses that propagate in water. This signal transmission by fish using electric impulses resembles the electric telegraph and Morse code, which were invented in the second half of the 19th century and used until radio and telephone replaced it.

Diogenes: I hope the evolution of electric fish doesn't go so far that fish start communicating with each other using built-in mobile phones.

Ralph: And it amazes me how fish, in the course of their evolution, managed to outpace humans in creating electric batteries. It is known that the first galvanic cell was invented by the Italian scientist Alessandro Volta in 1800. Yet rays, eels, and other electric fish have probably been using galvanic elements for hundreds of thousands of years.

Diogenes: What's so surprising about that? Nature is wiser than man.

Authors: We also think there's nothing surprising here. After all, all living organisms consist of cells. And a cell is an electric system. Every second, billions of potassium, sodium, chloride, and calcium ions pass through the cell membrane in both directions. In equilibrium, these ion flows establish a certain value of what is called the membrane potential. The membrane potential is the difference in electric potentials between the inner and outer sides of the cell membrane. Depending on the cell type, the membrane potential can vary from –10 mV to –100 mV. This is almost a ready-made microbattery. It is known that under the influence of the membrane potential, muscle cells change their size, resulting in muscle contraction. But how did nature manage to create an electric battery at the cellular level?

Scientists have found that if muscles are deprived of the ability to contract and change the conformation of proteins in the cell membrane, all ions passing through the membrane will go toward generating a positive charge. Such cells are called electrocytes—specialized cells capable of generating and transmitting an electric impulse.

By arranging electrocytes in a certain order and connecting them like batteries, in series, a sufficiently high voltage can be achieved. The body of an eel or ray contains millions of such batteries, constantly generating electricity. Scientists have sequenced the genome of electric fish. Comparative analysis has shown that the same genetic and cellular mechanisms are involved in the construction of the electric organ in different fish.

Reader: Yes, it's not as complicated as it seemed. There are tiny batteries—electrocytes. All that's left is to connect them correctly for freshwater and marine fish.

Authors: Well, in order to connect these batteries correctly, one must know Kirchhoff's laws and the formulas derived from them, which are provided in the Table 32.1. And one needs to have some understanding of electrical circuits.

Table 32.1 Electrical schemes of the ray and eel

Ray	Eel
$\mathcal{E}_{tot} \approx 20 - 60$ V, $I_{tot} \approx 50$ A	$\mathcal{E}_{tot} \approx 600 - 1200$ V, $I_{tot} \approx 1$ A

Electric eel as a tool for genetic engineering

To conclude our discussion on the abilities of electric fish, we have yet another astonishing thing to share. It turns out that the electric impulses of an eel also represent a genetic tool that allows surviving fish to acquire new abilities.

It appears that a shock from an electric eel can allow fish to acquire new DNA and new abilities. Scientists working in genetic engineering use the method of electroporation: creating temporary pores in cell membranes using a strong electric field. These temporary pores in the membrane are necessary to allow foreign DNA to penetrate the cell. To investigate whether such a phenomenon could occur in nature, scientists placed electric eels (*Electrophorus electricus*) and zebrafish larvae (*Danio rerio*) in an aquarium together with freely floating genes encoding green, fluorescent coloration. Incredibly, after two days of exposure to the electric field of the eels, some larvae began to glow green, indicating that their cells had accepted and started to express the foreign genes. These experiments raise a fundamental question: can fish or any other living organism in the natural environment acquire new genes in this manner and pass them on to their offspring? If so, electric impulses from eels could lead to new mutations that would affect the evolution of the species.

Reader: The example of electric eels clearly shows a new aspect of natural selection in the wild. Some fish perish under the impact of powerful electric field pulses, while others capture new genetic material, mutate, and give rise to new species.

Chapter 33

These Mysterious Cats: An Interview with a Renowned Philosopher

> *It all seems to me that I am just about to start living.*
> *Funny, isn't it? Cat's vitality!*
> —F. M. Dostoevsky

Journalist from a popular science magazine "Open Science": Dear cat, may I ask you a few scientific questions?

Diogenes: It's delightful to hear you addressing a cat so courteously. Cats are usually addressed informally, *"even though no cat has ever become drinking buddies (Bruderschaft) with anyone."*

Journalist: Oh, you're quoting Mikhail Bulgakov already.

Diogenes: Yes, every domestic cat has access to the home library and can read all the books there. Hence, I read at night. I've recently reread Fyodor Dostoevsky and found his novels to be "densely populated" with feline family members. So, what are your questions for me?

Journalist: Interestingly, according to statistics, the attachment to cats is not related to a person's age or gender but only to their deep inner need to love and care "just because," "for nothing," without expecting any gratitude in return, enjoying their purring. And here's the main question that has long puzzled scientists. Why do cats purr, what do they need it for, and how do they do it?

Shadowless Squids: Stories of Physics in Nature
Vitalii Zablotskii and Tatyana Polyakova
Copyright © 2025 Jenny Stanford Publishing Pte. Ltd.
ISBN 978-981-5129-43-4 (Hardcover), 978-1-003-57062-2 (eBook)
www.jennystanford.com

Diogenes: Your one question actually includes three: why, for what purpose, and how. Let me say right away that the answers to these questions lie in the realms of psychology, medicine, and physics.

Let's start with psychology. The purring or rumbling of a beloved cat has a calming effect on its owner. You could say that a purring cat offers its owner the best and absolutely harmless antidepressant. It is known that a cat usually purrs only when it feels safe, comfortable, and content. This is transmitted as low-frequency mechanical vibrations to the person who feels them by placing a hand on their beloved pet.

Journalist: Sorry for interrupting you. But I read in a scientific journal that a very strong magnetic field has an antidepressant effect on mice. Does that mean cats also create a magnetic field around themselves?

Diogenes: No. Cats emit low-frequency sounds accompanied by the vibration of their bodies. These are self-sustaining oscillations with a frequency of 20 to 30 Hz. And very importantly: purring does not necessarily require active muscle contractions.

From a physics perspective, much is unclear. Indeed, domestic cats are small animals, most of them weighing about 4 kg, and physicists are puzzled over how these animals manage to generate low-frequency vocalizations involved in purring. Such frequencies are usually observed only in much larger animals, such as elephants, which have much longer vocal cords. And while big cats, such as lions and tigers, can emit a loud roar, domestic cats can only produce low-frequency purring.

Scientists who studied the purring process of cats concluded that the sound produced by a purring cat is generated without muscle contractions or any brain signals. Once the vocal cords start vibrating, no additional neural input is required to sustain them.

Journalist: Is this purring on autopilot?

Diogenes: Yes, something like that. But the mechanism that excites these self-sustaining oscillations is not fully understood. Scientists have yet to find an answer to this mystery.

Journalist: So, cats purr in some unknown manner and for unknown reasons. But why do they do it?

Diogenes: To answer that question, I must inform you of some details from the field of medicine. It is known that low-frequency mechanical vibrations stimulate cell growth. Moreover, mechanical vibrations can change the direction of stem cell differentiation. In particular, under the influence of mechanical vibrations, stem cells are more likely to become bone cells rather than fat cells. Thus, purring and vibrations help cats heal minor bone injuries incurred during jumps or falls. You probably know that cats have more bones than humans, and this explains their extraordinary spinal flexibility. Purring and vibrations are a way to maintain and repair the skeleton and some internal organs. Perhaps this is related to cats' high vitality. Moreover, vibrations serve as a kind of auto-massage, promoting weight loss. Or, speaking scientifically, vibrations encourage stem cells to differentiate into bone cells, not fat cells. And of course, a beloved cat purring on its owner's lap brings true pleasure to the person and serves as the best antidepressant.

Journalist: Yes, I completely agree with you. When my cat purrs on my lap, I myself want to purr. And now I know that it would be beneficial. After all, stimulating the growth of stem cells is a very promising direction in regenerative medicine.

Diogenes: You're right. But scientists are just beginning to move toward regenerative medicine. And cats have known and used this for tens of thousands of years. Here's another example. Have you noticed how a cat jumps from the ground, e.g., onto a high windowsill? If not, pay attention. First, the cat looks at the spot it needs to jump to for a few seconds. At the same time, it calculates the necessary push force and determines the exact direction of the jump and its trajectory. And only then does it jump. But look, it jumps exactly to the height it needs. Not a centimeter more, not a centimeter less. This is an absolutely precise mathematical calculation, based on the laws of physics. Cats can easily jump to a height that is about ten times their height. If a human had the same ability, they could jump several meters up.

Journalist: Yes, that's true. The internet is currently abuzz with a mechanical dog stuffed with modern electronics, a multitude of sensors, and video cameras. It's still very expensive and not available for purchase. It can do everything: play with its owner and other dogs, follow various commands, run for 20 km without stopping, and even talk. But if it's tasked with jumping onto a windowsill, its computer brain freezes while solving such a complex mathematical problem.

Diogenes: You're right. As you might guess, I'm not fond of dogs, especially mechanical ones. But clearly, Newton's laws weren't programmed into its brain. I'll say more. This mechanical dog reminds me of a dreadful creature—the mechanical hound from Ray Bradbury's famous novel *Fahrenheit 451*. We live in a time when fiction becomes reality. And it's unclear whether that's unfortunate or fortunate.

Journalist: Why can't cats climb down from a tree, e.g., headfirst?

Diogenes: This is a matter of pure mechanics. Anatomically, cat claws are designed so that animals can only hook onto something when moving upward. A cat cannot climb down a tree headfirst. Hence, cats often find themselves in a predicament, having climbed a tree or scaled curtains. Incidentally, squirrels have more versatile claws, allowing them to descend a tree headfirst.

Journalist: Alright, we've somehow figured out the mechanics of cats. But our readers are also puzzled by the electrical properties of cats. Here's the question. My friend's cat often emits sparks in the dark. And I must say, it's very beautiful. But my cat, for some reason, cannot emit sparks, no matter how much I ask. Why?

Diogenes: This is simple electrostatics. You learned in school that many bodies become electrified by friction. For example, if you rub an ebonite rod against wool, the wool becomes positively charged and the rod negatively charged. When charges of opposite signs come close, an electrical discharge occurs—a brief electrical current. In the air, an electrical discharge is accompanied by sparks. A bright example of a large discharge in the air is lightning.

So, when you pet a cat or it rubs against something dielectric, its fur becomes electrified, i.e., acquires a positive charge. When the cat comes into contact with a negatively charged surface, an electrical discharge occurs, and you see sparks—mini-lightnings. Understand?

Journalist: Yes. But why can my friend's cat emit mini-lightnings, and mine cannot?

Diogenes: The thing is, the electrification process greatly depends on humidity. If the room's humidity is normal or quite high, fur electrification practically doesn't occur. Apparently, the humidity in your friend's apartment is quite low. Hence, his cat often becomes electrified and emits sparks.

Journalist: There's another question that many of our readers with domestic cats are curious about. Why do cats bring home the bodies of mice or birds they've caught?

Diogenes: The answer is very simple. By doing this, cats want to show that humans need to learn how to hunt. Then they will always have enough food regardless of supermarket prices. Moreover, natural food is always healthier than what we buy in supermarkets.

Journalist: It's hard to argue with your last statement. But since we've already started talking about hunting, I have one more question. Is it true that cats are the most efficient predators on the planet?

Diogenes: Yes, that is absolutely true. In 60% of cases, our hunts end successfully. And I think wolves, tigers, and lions, whose hunting successes are much more modest, envy us.

Journalist: Thank you for the interview. I think that at least a billion of our potential readers—cat owners—will gain new knowledge about their pets.

Diogenes: You're welcome. It was a pleasure to talk to you.

Chapter 34

Snakes: Hyper-sensitive and Insensitive

According to an ancient legend, during the dawn of life on our planet, when the first living organisms emerged from the ocean, Evolution was distributing legs to all comers. Centipedes were the first to arrive at the leg distribution point, and each took a hundred legs for themselves. Following them, octopuses arrived and received 8 legs each. Then came a crowd of various animals, and all received four legs. Probably, this was the optimal set for moving on land. The last legs were given to a monkey and a human, who received only two legs each. True, the monkey really wanted four legs, but Evolution's warehouse had run out of legs. Therefore, the monkey decided to use its hands as additional legs. Birds also received two small legs, but only from the last remnants of the leg supply.

Snakes were the last to arrive at the leg distribution point and asked for four legs each. But Evolution told them that it had no legs or wings left. "What should we do then?" the snakes asked, offended. "How are we supposed to move and hunt?" they added.

"I will teach you to crawl. It is a silent and versatile way of moving. And for hunting, I will give you two additional fangs filled with deadly venom," Evolution answered and added, "Also, I will give you thermal vision sensors—temperature sensors that will allow you to see prey in complete darkness. You will perceive infrared

Shadowless Squids: Stories of Physics in Nature
Vitalii Zablotskii and Tatyana Polyakova
Copyright © 2025 Jenny Stanford Publishing Pte. Ltd.
ISBN 978-981-5129-43-4 (Hardcover), 978-1-003-57062-2 (eBook)
www.jennystanford.com

radiation using special thermal eyes and special ion channels on the cell membrane. Take these additional thermal eyes while I still have them. And don't be upset, without legs, you will be pressing your entire body to the ground and feeling the slightest vibrations caused by the movement of potential prey. You will be hyper-sensitive organisms in my creation."

Snakes said, "Can we also have a stomach and mouth that allow us to swallow prey larger than ourselves?"

Evolution responded, "Yes, please, I give you such abilities if it compensates for the lack of legs."

Snakes thanked Evolution for its generosity and slithered away.

Reader: Something about those two additional fangs filled with venom is already scaring me. Is there a way to quickly tell if a person was bitten by a non-venomous or venomous snake?

Authors: Yes, you should look at the wound. If there are two deep small holes, it is a bite from a venomous snake. But if the wound looks more like a scratch with traces of small teeth, then it's a bite from a non-venomous snake.

A bit of snake physics. Any body that is warmer than its surroundings can be seen in infrared rays. Infrared light, by its nature, is the same light, only with a wavelength longer than that of visible light. Some snakes can distinguish bodies that are only 0.1 °C warmer than the surrounding environment. The pits located near the eyes of snakes are organs sensitive to thermal radiation. These holes have a diameter of about a millimeter and lead to a small cavity of about the same size. The walls of the cavity contain a membrane with a matrix of about 40×40 thermoreceptor cells. Unlike the rods and cones of the retina, these cells react not to the "brightness of light" of thermal rays but to the local temperature of their membrane. It's somewhat like the light-sensitive matrix of a digital camera. Only the sensors in the camera's matrix respond to the wavelength of light, not temperature, like in a snake.

Due to the large diameter of the thermal eye's entrance hole, the "thermal image" on the membrane turns out to be extremely blurry, but the snake manages to quickly reconstruct a fairly sharp picture of the surrounding world from it. The word "manages" conceals a complex process of mathematical processing of the blurred image to obtain a clearer picture. For example, in complete darkness, a snake sees a "blurry" mouse but quickly recalculates this image into a clear one with its brain and unerringly attacks it. Snakes' thermal hypersensitivity is combined with quick response, allowing snakes to react to the thermal image of prey in less than 35 ms. Snakes' ability to sense heat is so great that they can detect the heat emitted by a rat from a significant distance. Snakes can hunt at night and pursue their main prey—rodents in their underground burrows.

Boas and pythons, which also have thermosensitive organs, instead of thermal pits, have more than 13 pairs of thermoreceptors located around the lips.

Ralph: Now I have a question for Diogenes. Can cats catch mice in complete darkness? In other words, do cats have thermal sensors that allow them to see a blurry mouse in the dark?

Diogenes: Dear dog Ralph, cats, like other warm-blooded animals, cannot have thermal receptors. Do you know why?

Authors: Allow us to answer the question "why?" Let's consider this issue from a physics perspective. Let's start with the fact that every heated body emits electromagnetic waves. If the body's temperature is within the range of room temperatures or slightly below or above, such bodies emit electromagnetic waves with a wavelength falling in the infrared range. Now imagine that the temperature sensor is located on the body of a warm-blooded predator. What radiation does this sensor receive? Correct, it registers infrared waves emitted by the body of the animal itself. And somewhere far away is prey, e.g., a mouse, which is also warm and emits electromagnetic waves in the same wavelength range—infrared, invisible to us light. And these waves also reach the predator's thermal sensor. So, two waves fall on the sensor: from the distant mouse and from the predator's body. Can this sensor register the thermal waves that came from the mouse?

Answer: theoretically—yes, but practically—no. The reason is that the intensity of the waves (the energy carried by the wave per unit of time through a unit area) rapidly decreases with distance. For a point source of waves, the intensity decreases inversely proportional to the square of the distance from the source. What does this tell us? Correct, that the intensity of the waves reaching the sensor from the mouse will be much less than the intensity of the waves coming from the body of the predator itself, i.e., its thermal noise. As a result, the sensor will register a signal from the predator itself but not from the prey. In physics, in such cases, it is said that the signal-to-noise ratio is much less than one, and this makes it impossible to register a useful signal and speaks of the uselessness of thermal sensors on the body of warm-blooded animals. However, for cold-blooded animals, such a problem does not exist. The surface temperature of a snake's body is the same as the temperature of the ground or stones it lies on. Not for nothing do people say: cold and insensitive as a snake. All snakes without exception are predators.

Figure 34.1 Vision in infrared.

Diogenes: I have a riddle for the reader. There was such a curious incident. Two young pythons were fighting over prey—a rat. When the rat ran away, one python simply swallowed the other.

And now the question: "If two identical pythons start swallowing each other from the tails, how will it all end?"

Reader: I have to model this situation on a computer. After all, I learned the Python programming language.

In conclusion, here are a few more interesting facts about snakes.

The lower jaws of pythons are attached to the skull in such a way that they allow them to open their mouths wide to swallow large prey. Residents of the Indian state of Gujarat witnessed an unusual feast of a huge python, which did not end well for the snake. The python attacked an antelope and swallowed it whole. But it overestimated its strength. The snake couldn't even move. After spending several minutes in convulsions, the python died.

Currently, there are known to be three and a half thousand different species of snakes.

The most venomous terrestrial snake in the world is the taipan, which has venomous teeth up to 13 mm long. A single dose of taipan venom is enough to kill a hundred people. This snake is 50 times more venomous than a cobra. The taipan is also considered the most dangerous snake, as its super venomousness is combined with huge size (up to 3.5 m in length) and an extremely aggressive nature. The longest venomous fangs are possessed by the Gaboon viper, their length can exceed 40 mm.

Snake digestive enzymes dissolve everything except the victim's fur, feathers, and claws. The snake's heart has an interesting feature. While in humans and other animals it is fixed in one position and in one place, in snakes, it can move throughout the body. Why? If the reptile has swallowed food, the heart can retract to the side to let the prey pass further and then return to its original place.

Snakes lack an external and middle ear, as well as eardrums. However, they are very sensitive to vibrations, so they can capture sound waves traveling through the ground. They "hear" them with their entire body, as there are mechanical vibration sensors on the belly of a snake. They allow capturing any vibration of the Earth's surface.

Chapter 35

The Seventh Sense and Earthquakes

Early in the morning of April 26, a strong earthquake occurred in the city of Tashkent. As a result of the earthquake, the central part of Tashkent, a city with a population of millions, was almost destroyed. Several days before this, the city authorities were approached by elders—old and wise highlanders—who said that a strong earthquake was imminent. When the city's mayor asked how they knew this, the elders replied that snakes and lizards had left their burrows, sensing the approaching earthquake. And they were absolutely right. But the city authorities did not listen to them.

Here's another example when people listened to a dog and survived during a strong earthquake in Turkey in 2023. A family lived on the outskirts of a village. A strong earthquake started at night. People ran out of the house and dashed toward the road in the darkness. But their dog suddenly started pushing the people away from the road with all its might. It was night and visibility was poor, but what they saw a moment later shocked them: the road began to collapse. A chasm formed right in front of them. Terrified, people ran in the opposite direction. But the dog suddenly grabbed its owner by the sleeve and started dragging her away. It turned out that a new deep crack had formed exactly on the side where the frightened people were planning to run. This happened several times because the dog kept people away from where the crack was about to appear.

Shadowless Squids: Stories of Physics in Nature
Vitalii Zablotskii and Tatyana Polyakova
Copyright © 2025 Jenny Stanford Publishing Pte. Ltd.
ISBN 978-981-5129-43-4 (Hardcover), 978-1-003-57062-2 (eBook)
www.jennystanford.com

In the morning, when everything had calmed down, the woman saw many holes in the sleeve of her clothes from the dog's teeth. But everyone was happy and thanked God and their beloved dog for the rescue.

Undoubtedly, some animals and amphibians can sense the approach of strong earthquakes in advance. But, as is known, to predict anything, one must see or feel changes in the environment or natural phenomena. And these changes must be rapid, as slowly changing external conditions cannot be correctly perceived by living beings. This is best illustrated by the so-called boiled frog effect. Here's the thing: if you put a frog in a large pot of water and very slowly heat it up, the frog swimming in the water won't notice the change in water temperature until it dies in the boiling water. But if you throw a frog into hot water, it will instantly jump out of the pot. Feel the difference? Slowly changing conditions over a short observation time do not allow for any prediction. But if some physical quantity starts changing rapidly, it's already a signal that carries certain information for a living being. Now let's move on to changes in the geomagnetic field, which can be interpreted as signals—harbingers of earthquakes. It's known that the Earth's magnetic field is subject to quite strong fluctuations. What could cause these fluctuations? The geomagnetic field is generated by at least two sources. The first is the Earth, as a large magnet, and the second, relatively weaker, is streams of charged particles (ions) in the upper layers of the atmosphere.

Disturbances in the magnetic field before earthquakes

First, let's consider the changes in the field generated by the second source—the ionosphere. Daily changes in Earth's atmospheric temperature lead to corresponding changes in ion flows, and, consequently, changes in Earth's magnetic field. As a result, the Earth's magnetic field experiences fluctuations with an amplitude of 30–60 nT, which is about 0.1% of the average value of the Earth's magnetic field induction (50 μT). But are rapid temporal changes in the magnetic field from the main source possible? Probably not, if one crudely imagines it as a huge magnet buried in the ground. But that's not the case. As we already know, the sources of the geomagnetic field are giant currents flowing inside our planet in the form of charged fluids. Therefore, any disturbance in their movement can lead to changes in the geomagnetic field. Before the main shock of an earthquake, there is a certain preparatory period, during which layers of the Earth's crust may undergo deformation, break down due to increasing pressure, become electrified due to friction, etc. This is not yet an earthquake, but the destruction and deformation of rocks have already begun, leading to the release of relatively small mechanical and electromagnetic energy. Directly before the earthquake, the states of the Earth's crust and mantle are not in equilibrium. They are unstable, including in the mechanical sense, i.e., somewhere pressure increases and the shifting of Earth's plates to a more equilibrium position begins a few days or hours before the earthquake. Obviously, before a strong earthquake, such shifting can locally change the circulation of currents (flows of charged fluid) that create the main magnetic field of the Earth. This hypothesis is indirectly confirmed by experiments, which will be discussed below.

Searching for magnetic precursors to earthquakes

Let's say right away that the mechanisms for exciting magnetic precursors to earthquakes are not reliably known. The method of exciting electrical currents in the Earth's crust during the preparation phase of an earthquake is, in particular, unknown. Nonetheless, it has been experimentally established that, on average, the amplitude of magnetic field variations before strong earthquakes is higher than after them. For instance, a rather sharp change in the amplitude of variations in the averaged perpendicular component of the magnetic field was registered before and after strong earthquakes: 5 nT/min 12 h before the earthquake and 2 nT/min 12 h after it. Therefore, when talking about magnetic precursors to earthquakes, we must consider possible changes in the physical state of both sources: (1) variations in the magnetic field caused by changes in the ionosphere's state; (2) disturbances in the geomagnetic field even in areas quite distant from the epicenter of the future earthquake in the preceding period. The preparatory phase of an earthquake, during which the accumulation of elastic energy of Earth's plates gradually occurs, can last hundreds and thousands of years. During this, no noticeable signals emanate from the nascent earthquake's epicenter, and no changes in the magnetic field are observed around it. Then this slow phase is replaced by a short-lived critical phase, in which processes of slow displacement of Earth's plates already begin. During this time, small physical signals (acoustic, electromagnetic, and seismic) come from the epicenter of the maturing earthquake, and they can be registered by both physical instruments and the senses of certain animals. As the moment of the earthquake approaches—approximately a day before it—disturbances in the magnetic field with maximum amplitude begin to occur.

Figure 35.1 Famous scientist.

Do animals register and understand magnetic signals preceding an earthquake?

Instances where animals or birds, through their unusual behavior, predicted the onset of an earthquake have been recorded both in the distant past and in our times. Living beings react differently to the impending seismic event: they either start to show increased anxiety, run around chaotically, make alarming sounds, or suddenly become quiet, as if listening to what is happening inside the Earth's bowels. Scientists have managed to determine the time when animals begin to react to changes in Earth's electromagnetic field. It was noted that animals rarely manage to predict an earthquake more than a day in advance. But a day is precisely the characteristic time before the onset of an earthquake when the greatest changes in Earth's magnetic field occur. Of course, there are several other possible reasons for the unusual behavior of animals before seismic shocks. One of them could be that the movement of underground plates before an earthquake creates electrical charges and the corresponding electric field, which some animals can sense. Another hypothesis suggests that animals can feel even very weak shocks of the future devastating earthquake, unnoticed by humans. Indeed, seismic (acoustic) waves spread faster and better through the Earth's crust than through the air. And obviously, animals and reptiles living in burrows can be the first to feel these weak shocks and sounds. Moreover, according to scientists, reptiles and amphibians sense minor changes in the chemical composition of water and soil that occur on the eve of shocks due to partial destruction of rocks. In other words, before an earthquake, the composition of underground and soil waters changes, which can first be sensed by those living in burrows and in water.

Studying animal behavior before earthquakes is quite difficult for scientists, as large earthquakes are relatively rare, and seismologists predict them with varying success. But there are countless facts that testify to animals predicting earthquakes. Changes in animal behavior before earthquakes were noted as far back as 328 BC. One of the ancient Greek thinkers wrote that a few days before the earthquake that destroyed the city of Helike, moles, weasels, echidnas, and centipedes emerged from their burrows and turned to (as it seemed to him) chaotic escape. To date, through observations, it has been established that animals are capable of predicting earthquakes with a magnitude of four and above. Meanwhile, only those animals that are within a 100 km radius of the epicenter can feel the approaching natural disaster. Since humans observe domestic animals much more frequently, it is believed that dogs, cats, horses, sheep, domestic birds, and amphibians react faster than others to the approach of disaster. Numerous examples of how animals, with their earthquake predictions, saved a vast number of human lives can easily be found on the internet.

Reader: From this story, I concluded that modern science, despite all its advanced and ultra-sensitive measuring devices, cannot predict earthquakes in advance and reliably. Is this true?

Authors: Yes, you are right. Long-term forecasting of earthquakes remains an unsolved problem for modern science. Therefore, just as thousands of years ago, today cats and dogs are the main predictors of earthquakes. Anomalous behavior of cats and dogs has been recorded in almost any seismically active zone of the world. Some newspapers and online publications make big sensations out of this, claiming, e.g., that a cat or dog saved hundreds of thousands of city residents. Of course, one should not believe such sensations, but one should unconditionally believe their cat or dog.

A scientific article was recently published in which, based on numerous observations, an intriguing correlation was found between earthquakes and cosmic radiation. Analyses were conducted using several statistical methods and sensors installed almost all over the surface of our planet. In each case, over the study period, a clear correlation was identified between changes in the intensity of secondary cosmic radiation and the sum of magnitudes of all earthquakes with a magnitude greater or equal to 4. This correlation shows that earthquakes with a magnitude above 4 occur exactly 15 days after a registered change in the level of solar radiation. This is good news, as it suggests the possibility of advance prediction of impending earthquakes.

The explanation of this correlation is based on two facts. First, Earth's magnetic field deflects the trajectories of charged particles of cosmic radiation—solar wind. Second, the generation of Earth's magnetic field is conditioned by the movement of flows of charged fluid (electric currents) in our planet's liquid core. So, before an earthquake, minor shifts in tectonic plates introduce disturbances into the movement of material flows (electric currents) generating the geomagnetic field. Thus, in the vicinity of an impending earthquake, Earth's magnetic field changes, leading to a change in the trajectories of charged cosmic particles over this area. In other words, the intensity of cosmic radiation, registered over the future earthquake site, depends on the dynamics of disturbances occurring inside our planet.

Chapter 36

About the Senses of Sharks

No, no, we won't be talking about the feeling of love between a female shark and a male shark. Nor will we discuss a shark's sense of compassion toward its prey. Everything will be much simpler. And, after all, it's a bit strange to discuss sharks in a scientific book. A shark is more likely a character for adventure novels from the time of Pirate Drake. But nonetheless, it is precisely the shark that stands as one of the unique living organisms that utilize both electrical and magnetic phenomena in its adventurous life. This predatory machine is equipped with high-strength mechanics (its teeth are regularly replaced as they fall out, following a conveyor belt principle—their replacements continuously grow from the inside; an adult great white shark can bite through a steel rod as thick as a little finger), high-precision magnetic navigation, and ultra-sensitive electrical sensors. Moreover, sharks have especially well-developed photo and chemical reception sensors. Not every modern submarine can boast such an arsenal. But we don't intend to scare you further, so let's move on to the senses of sharks.

It's known that some shark species embark on oceanic journeys spanning thousands of kilometers each year. For instance, it was recorded that one great white shark regularly traveled from South Africa to Australia and back. But how it navigated the ocean remained a mystery for a long time. Finally, in 2021, scientists

Shadowless Squids: Stories of Physics in Nature
Vitalii Zablotskii and Tatyana Polyakova
Copyright © 2025 Jenny Stanford Publishing Pte. Ltd.
ISBN 978-981-5129-43-4 (Hardcover), 978-1-003-57062-2 (eBook)
www.jennystanford.com

were able to prove that sharks can sense both the direction and magnitude of the magnetic field and use Earth's magnetic field map for orientation during their long wanderings. Earth's magnetic field is heterogeneous, and on the path of sharks' long migrations, the magnitude of the magnetic field changes from 27 µT to 56 µT. For example, the magnitude of the magnetic field in the Gulf of Mexico is less than that 600 km north of the continental part of the USA. And probably, sharks are well aware of this.

It's very likely that sharks have an accurate map of Earth's magnetic field distribution in their heads. Otherwise, how would sharks return to the estuaries of the same rivers—to their homes, after such long journeys? Everything suggests that sharks precisely know where their home is and can flawlessly find their way back, using Earth's magnetic field map. To prove the fact that sharks use the magnetic field for navigation, scientists measured the magnetic field in their habitual habitats, and then, using an electromagnet, created exactly the same magnetic field in terms of magnitude and direction in a large circular pool. Under these conditions, sharks always tended to swim in the direction of the magnetic field created by the electromagnet. These experiments proved that sharks can orient themselves using the magnetic field, and most interestingly, they have a rather detailed magnetic map in their heads. However, it's unclear whether this magnetic map was inherited from their parents or if they recorded it in their memory during their first journey.

Furthermore, we still don't know how and with what sharks sense the magnetic field. One hypothesis—magnetoreception through electromagnetic induction—suggests that a shark senses the electric field, which is induced in its special conducting lateral lines during the shark's rapid movement in Earth's magnetic field. Indeed, according to Faraday's law, an electric field is induced in a conductor moving in a magnetic field, the magnitude of which is proportional to the rate of change of the magnetic flux. As is known, some shark species can reach speeds of up to 50 km/h. Thus, according to this hypothesis, any living organism with the ability to sense an electric field potentially can feel a magnetic field too, if its movement speed and the intensity of the magnetic field allow inducing a detectable electric field in its body. Incidentally, from a biological standpoint, sharks' closest relative is the electric ray.

Reader: Oh, now I know how to catch a shark! I need to create a magnetic field with magnets, just like at their home, and they'll swim over for a visit, right?

Authors: In principle, you're correct. But don't forget that sharks have many other sensors. For example, sharks can sense very small electric fields with a strength of only 0.01 μV/cm = 10^{-6} V/m. This allows them to detect prey by the electric fields generated by the respiratory muscles and heart of the prey. The shark's electroreceptive apparatus is represented by the so-called ampullae of Lorenzini—small capsules embedded in the skin with tubes extending to the skin's surface. Inside the skin, they end in bundles of cells that can perceive electric fields through electroreceptors—cells on whose membranes are special ion channels that can open or close depending on the magnitude of the electric voltage. Finally, these cells transmit this information to the nervous system by releasing packets of chemical messengers called neurotransmitters, which reach synapses and points of connection with adjacent neurons. Studying shark behavior in electric fields, scientists discovered that when the electric field was turned on, the breathing rate of sharks increased to the level observed at the smell of food. Do you know what this indicates? Yes, correctly, this fact indicates that the entire electrical system of a shark is tuned for one thing: to detect and attack prey!

It is also assumed that the ampullae of Lorenzini are used by sharks as thermosensors, determining the temperature of the environment with an accuracy of up to 0.05 °C. However, it's unclear why a shark needs to measure water temperature with such high precision. Maybe because most sharks are cold-blooded animals, i.e., their body temperature is equal to the temperature of the surrounding environment. But unlike most sharks, great white sharks are partially warm-blooded, allowing them to move faster during a chase. Why do you think warm-blooded sharks swim faster than cold-blooded ones? Note that on average, an adult great white shark consumes about 11 tons of food annually.

Sharks have a very keen sense of smell: they can detect blood diluted in seawater in a proportion of one to a million. Just imagine: one molecule of blood per million molecules of water! Interesting, how do they manage to register one molecule among a million others? Sharks never sleep; otherwise, they would simply suffocate.

The thing is, they can pass water through their gills, extracting oxygen from it, only while moving, as shark gills in a state of rest cannot "suck in" water on their own. So, they live in constant motion, not so much to travel and admire the underwater views of tropical coral reefs or enjoy the beauty of Arctic seas, but simply to breathe and sense electric and magnetic fields. Incidentally, having a map of Earth's magnetic field and an internal compass, sharks make long journeys practically in a straight line, just like an airplane in the sky.

So, a shark has a well-developed "sixth sense"—the ability to perceive electromagnetic fields. The shark's electrical "sixth sense" is tuned for attack, while the magnetic "sixth sense" is tuned for navigation. But we cannot claim to know everything about the senses of sharks yet. Studying sharks is not only a way to gain knowledge that can be applied in human practice but also a fascinating way to learn how evolution shapes senses. But, as you can guess, conducting experiments with sharks is quite difficult and sometimes dangerous.

Diogenes: Oh, no, I'm not planning on catching sharks, although I do love eating fish. I want to ask, how are sharks doing with their knowledge of physics?

Authors: Their understanding of physics is just fine. For example, the thresher shark (fox shark) has a very long upper lobe of its tail fin. And it's not for better swimming. Zoologists have discovered that the enormous fin is used by sea foxes for hunting. During hunting, they can stun fish with it or herd them into schools to catch more. But the interesting physics is in its tail.

A sea fox can hit fish with its tail so powerfully that the maximum tail movement speed reaches 22 m/s (=79.2 km/h)! Such speed in water creates small bubbles due to cavitation. Scientists have found that through evolution, thresher sharks have developed a spine that works on the principle of a trebuchet—a throwing device using the force of gravity and a massive counterweight. And the cartilaginous skeleton of sharks, which lacks bones, functions as a "living spring" that accumulates mechanical elastic energy and allows for very fast and accurate tail strikes.

It's interesting that in the matter of hunting perfection, hammerhead sharks took a different path: unlike thresher sharks, they didn't elongate their tails but expanded their heads. The head of a hammerhead shark looks as if its skull has been stretched out to the sides. Because of this, the fish's head has come to resemble a hammer, and the body looks like the handle of a hammer. With a length of 6 m, the distance between the shark's eyes can reach 1 m. Evolution worked for millions of years to give the shark such an appearance, trying to adapt it as much as possible to its habitat and way of life.

Ralph: And what advantages does such a wide head give to the shark?

Authors: Firstly, the hammer-shaped head provides the shark with a very wide angle of view, reaching 360°. The predator sees everything happening in front, behind, above, and below. Nothing escapes its gaze, making the shark an excellent hunter. Secondly, the wide head serves as a large radar because it has more room for the ampullae of Lorenzini, which allow it to sense electrical fields. In hammerhead sharks, the ampullae are located on the underside of the head, enabling them to detect fish buried in the sand. Finally, scientists suggest that this strange head shape helps the shark sense the Earth's magnetic field. These sharks migrate long distances, and therefore, such a large compass is necessary for them.

Chapter 37

The Shadowless Squid

It is better to aspire to knowledge than to wealth.
—Omar Khayyam

But on the other hand:
Woe from Wit.
—Alexander Griboedov

Here are two opposing statements from great writers of the past. What to choose?

Can we even choose? It seems that the drive to acquire knowledge about the surrounding world and to pass it on to subsequent generations is embedded in nature itself and hidden somewhere in our genes. In one of the stories, did we talk about dark matter in DNA—the undecoded segments of the genome? No one knows what information is recorded there. Maybe it contains secrets about the structure of our Universe? Perhaps these genome segments program us to explore this world, to accumulate and enhance our knowledge about nature.

Reader: No, wait a minute. You think nature has programmed us to study it? Why? Why, e.g., does a squid need to know that there is a moon and how brightly it shines at night? Why does this squid need to know, e.g., about geometrical optics or electricity?

Shadowless Squids: Stories of Physics in Nature
Vitalii Zablotskii and Tatyana Polyakova
Copyright © 2025 Jenny Stanford Publishing Pte. Ltd.
ISBN 978-981-5129-43-4 (Hardcover), 978-1-003-57062-2 (eBook)
www.jennystanford.com

Authors: Thank you for the interesting question. Let's try to explain this using the squid as an example. The short answer: the squid needs to study physics because knowledge about nature helps it survive. And physics is the science of nature.

Figure 37.1 What do you choose: all the wisdom of timeless books or boundless wealth?

Let's start with a question. Can you quietly fly low above the sand on a clear moonlit night without casting a shadow?

You might answer, "I can't fly on moonlit nights, with or without a shadow. And why would I need to?"

Imagine that you've turned into a Hawaiian bobtail squid living near the shore in Hawaii in shallow knee-deep water. Note, this squid is nocturnal: it buries itself in the sand and sleeps during the day, and hunts at night. And nights in these places can be very romantic and bright.

On a moonlit night, when light from the stars and the Moon penetrates the water, any body moving in the water will be accompanied by its shadow gliding across the sandy bottom. What kind of hunting is that if the prey sees the predator's shadow moments before the attack? It's like when a cloud's shadow suddenly rushes over you, although with the sun's rays slanting, the cloud itself may still be far from above you and rain may not start anytime soon. Hence, you have time to find shelter. Similarly, the squid's prey sees its approaching shadow and immediately hides in the sand.

How then to rid oneself of one's shadow?

If you don't know a suitable method, you'll have to sit hungry and wait for bad gloomy weather when there are practically no shadows. Let's just say, hunting only in bad weather is actually a very bad option. Because those you intend to hunt also prefer not to leave their shelters in bad weather. So, you need to come up with something that allows you to glide near the bottom surface without casting any shadow even under a full moon: learn to glide above the bottom without any shadow. How to do it? Did you figure it out? No. But the squid did. Now we'll tell you how it does it. Don't be upset because you couldn't solve this task. Probably, you had too little time, while the squid had millions of years to find its solution. And the squid's motivation was clearly stronger than yours. After all, instead of solving such a task, you can simply open the refrigerator and move on to an easier task.

So, on bright nights when there's a lot of light from the moon and stars, this light penetrates the water all the way to the bottom where the squid lives, as the water depth here is about half a meter. And how does our squid solve the survival challenge? All very simple and simultaneously very high-tech. The squid's belly has plates that open or close a specialized light organ where bioluminescent bacteria live. These bacteria, typically found in the ocean and called *Vibrio fischeri*, have rented a part of the squid's belly (hopefully, at a reasonable price). This bacterium has a special property: it produces a glow called bioluminescence. It's similar to how fireflies glow. Moreover, the squid has light detectors on its back and can feel how much moonlight or starlight hits its back. It opens or closes the plates so that the light flow emitted by the bacteria and falling on the sand precisely matches the light flow hitting the squid's back from the moon and stars. The light coming from its belly illuminates the area of its shadow, and as a result, the squid does not leave any shadow!

Figure 37.2 The shadowless squid.

But that's not all the squid has learned. Our squid has learned to synchronize its biological clock with the bacteria's clock. You probably know that all living organisms have special genes whose task is to synchronize the internal daily rhythm with external conditions. These genes encode special proteins that accept and "recognize" visual data about whether it's day or night outside, and accordingly adjust the pace of the internal clock. Our squid has two such genes, encoding cryptochromes (CRY). One of the cryptochromes is located and works closer to the central nervous system, the brain, and the sensory organs. The other protein is synthesized in the squid's bioluminescent organs, where the bioluminescent bacteria live. Moreover, the activity of the second gene (escry1) matches the cycle of bacterial bioluminescence. So, the squid begins to glow with the onset of night when it goes hunting. Thus, the biological clocks of the squid *E. scolopes* and its bacteria are fully synchronized.

Reader: And I'm not surprised. Our gut microbiota already influences everything from immunity to pregnancy; why couldn't it also affect our biological clocks?

Authors: Yes, it might be so. But in the case of the squid, the bacteria live outside, not in the gut like in humans. Therefore, controlling them is more complicated. Some of our readers might know how, in physics, clocks can be synchronized using a light signal (a method proposed by Einstein). But as you can see from the squid example, living nature has solved this issue differently.

Why do we strive for knowledge?

Using the squid as an example, we see how knowledge helps it hunt successfully and survive. The squid knows a bit of astronomy (at least something about the Moon, maybe even about the stars), knows biology (about bioluminescent bacteria), uses the bacteria's glow (synchronizing its biological rhythm with the rhythm of symbiotic bacteria), and knows geometrical optics, directing the bacteria's light to the bottom in such a way that the prey doesn't see its approaching shadow and doesn't have time to hide. Based on this knowledge, the squid has developed its technology for becoming the ocean's stealth aircraft.

It can be asserted that all living organisms produce knowledge, accumulate knowledge, and transmit knowledge to subsequent generations.

Diogenes: The pursuit of knowledge is an evolutionary trait of life.

Reader: Maybe this is the main difference between living and non-living nature.

Ralph: I agree, for life, all beings need knowledge of physics, biology, and many other sciences. But why do we need knowledge about the Universe?

Authors: Let's think more broadly, and the answer will come by itself. Let's recall the most important knowledge humanity has acquired, say, in the last couple of hundred years. Above all, we see tremendous progress in physics, chemistry, biology, and medicine. From the steam engine to the thermonuclear reactor, from alchemy to quantum chemistry, from the discovery of bacteria to genetics, from homeopathy to gene therapy. And now imagine human knowledge, say, in hundreds of thousands of years. Obviously, human needs for knowledge and energy sources will increase hundreds and thousands of times. For example, suppose our Sun suddenly cools down (we still don't fully understand the processes occurring inside stars), and we have to look for a new warm star for comfortable living.

Reader: Maybe the dark segments of our genome contain secret knowledge about the structure of the Universe and molecular survival mechanisms in case of cosmic catastrophes?

Authors: Undoubtedly, in case of cosmic catastrophes, humanity will need knowledge about the Universe. Maybe something about the Universe is recorded in the dark parts of the genome. But we don't know that yet. Therefore, humanity is now accumulating knowledge at such rapid rates and in such large volumes that it no longer manages to record and transmit them at the genetic level. That's why people write textbooks, books, and place their knowledge on other material carriers: electronic, magnetic, and optical.

Reader: Ah, it would be nice if all knowledge was transmitted at the genetic level from generation to generation! Then I wouldn't need to go to school and learn all the subjects there. I think the knowledge accumulated by my parents, as well as my grandparents, would be quite enough for me to immediately pass the university entrance exams for a master's degree.

Ralph: Ah, if only it were so. Then my puppies would be the smartest puppies in the world!

Diogenes: Once, a very beautiful lady approached Einstein and said, "If we had a child, it would inherit my beauty and your intelligence." "And what if the opposite?" replied Einstein.

Authors: We've had a little fun, but now seriously. All living beings learn and accumulate knowledge, from the simplest bacteria and lower worms to higher organisms and humans. For instance, it is known that German Shepherds living under the same roof with humans can remember up to 200 words, while dogs living in the yard know only about a dozen words.

Diogenes: So what, dogs can learn 200 words. We, cats, understand absolutely everything, including the structure of the Universe. We just don't like to reveal our knowledge. Who knows how humans might start using it.

Ralph: I cannot agree that dogs can only remember about 200 words. We, dogs, understand absolutely everything, not just words, but also feelings. For example, a dog I know is currently working on a book *The Rights of the Dog and the Duties of Its Owner*. And as far as I know about cats, every second cat among you is a philosopher.

Diogenes: Well... not without it. After all, the word philosophy literally means the love of wisdom. And as it is written in the proverbs of King Solomon: "Wise men store up knowledge...." That's what we do and tell no one.

Authors: Somehow, we've moved from the squid to philosophical problems. But this topic is endless, like our Universe, which has existed for approximately 13–14 billion years and stretches from unimaginably small scales (comparable to the sizes of quarks and leptons or even the Planck length, $\approx 1.6 \times 10^{-35}$ m) to the sizes of galaxy clusters (tens of millions of light-years, $\approx 10^{23}$ m)! As you can see, we are faced with a world full of mysteries, opening up vast opportunities for acquiring new knowledge. But the most interesting thing is that we still **Do Not Know** whether a finite or infinite number of physical laws governs our Universe. And we very much hope that one of our young readers will find answers to this and other questions of the universe in the near future.

But to bring the reader back from cosmic space to Earth, we will ask him one question. What's bigger, a cloud or its shadow on the Earth's surface?

Reader: Since the Sun can be considered an infinitely distant source, its rays falling on the Earth are parallel. This means that the size of the shadow is practically equal to the size of the cloud. It's simply a parallel projection of the object onto a plane.

Authors: Excellent. This is precisely the fact of geometric optics that our squid uses, choosing the number of bacteria necessary to illuminate its shadow.

Chapter 38

Electric Car versus Sperm Cell

Sherlock Holmes: I received a call from Scotland Yard informing me that the criminal left a gentleman's kit at the crime scene. Dr. Watson, do you know what a gentleman's kit is?

Yes, Sir. A gentleman's kit—it's 23 chromosomes and a nanomotor.

"Would you like to investigate this crime, Ralph?"

Ralph: "No, today I'd rather revisit some quantum mechanics sections. Perhaps Diogenes would take on the investigation?"

Diogenes: "Detective cats! That's a novel term in criminology. No, thank you. I'm not a detective; I'm a philosopher."

Authors: "Diogenes and Ralph, that's enough joking around. Let's move on to discussing global issues.

While humanity struggles with internal combustion engines, promising to replace them with electric motors in the coming decades, nature, through evolution, has been using electric motors for hundreds of millions of years. Moreover, nature has made these electric motors so miniature and environmentally friendly that they fit perfectly into the design of many unicellular organisms: bacteria and sperm cells. Let's start with the structure of a sperm cell. What is it? It's a male reproductive cell, whose purpose is to deliver genetic information to the female egg cell. Sperm cells were first described by the Dutch scientist Antoni van Leeuwenhoek in 1677. The typical structure of a sperm cell reflects the form of the common ancestor of

Shadowless Squids: Stories of Physics in Nature
Vitalii Zablotskii and Tatyana Polyakova
Copyright © 2025 Jenny Stanford Publishing Pte. Ltd.
ISBN 978-981-5129-43-4 (Hardcover), 978-1-003-57062-2 (eBook)
www.jennystanford.com

animals: a unicellular organism with hardly any cytoplasm, moving by means of a flagellum at the back, using it as a propeller, or more accurately, as a rudder. The structure of a sperm cell fully matches its purpose in life—to deliver genetic information (chromosomes) precisely to the destination.

Isaac Asimov, a renowned science fiction writer, described the structure of a sperm cell best: it's like a kamikaze, having only a motor, a fuel reserve just enough for a one-way trip, and a light shell with tightly packed cargo that needs to be delivered. And nothing unnecessary. But no. Perhaps in addition to all the listed features, a sperm cell also possesses some form of minimal intelligence, enabling it to actively seek its target and navigate around obstacles on its way.

Nature has thus ordained: 23 chromosomes in a shell, a motor, a minimal fuel reserve, and minimal intelligence. And two more words on intelligence. A single sperm cell carries 37.5 Mb of information, encoded in DNA. It can be said that the sperm cell is the pinnacle of versatility, standardization, and efficiency of living nature.

In the human body, the sperm cell is the smallest cell: 5.0 μm long plus a flagellum (tail) 45 μm long, which helps the sperm cell move. During movement, the sperm cell usually rotates around its axis. Think about why? The movement speed of a human sperm cell can reach 0.1 mm/s or over 30 cm/h. In the female reproductive tract, sperm movement is against the flow of fluid. That's what the powerful motor is for! To achieve fertilization, the sperm cell needs to overcome a distance of about 20 cm. This is exactly what its energy reserve is calculated for. But what is this energy source, and how does the motor that drives its propeller work?

Reader: Wait! I have one question. You said think about.... I thought, but couldn't find an answer. But why does the sperm cell rotate around its axis during movement?

Authors: The answer to your question is the same as the answers to similar questions: Why does a cat always land on its feet when falling? Or why does a helicopter with a single rotor need another side rotor?

Reader: Oh, I'm very interested in knowing that. I have a cat, but it doesn't tell me how it manages to land on its feet every time it falls.

Authors: Well, since your cat is so uncooperative, we'll have to divert from the main topic, from nanoelectric motors. So, you've surely heard about the law of conservation of angular momentum, which states: the total angular momentum of a closed system is constant.

Reader: No, we didn't study this law in school.

Authors: But we're sure you've studied the law of conservation of momentum. Let's start with that and then move on to the law of conservation of angular momentum. Let's illustrate how the law of conservation of momentum works with a simple example. Imagine you're standing on slippery ice and forcefully throw a basketball straight away from you. What happens to your body? Correct, your body receives momentum in the direct opposite direction, i.e., you'll slide a bit backward, while the ball received momentum moving forward. The total momentum of the system (you + ball) is conserved and remains equal to zero. A very similar thing happens when one part of the system starts rotating relative to another: the total angular momentum of the system remains equal to zero. This means that if one part of the system begins to rotate, say clockwise, the remaining part will start rotating counterclockwise.

Now back to the falling cat. Unlike some readers, cats have learned the law of conservation of angular momentum.

Let's assume the cat accidentally falls from a height and initially during the fall, it is upside down and not rotating around its axis. This means its initial angular momentum was zero. According to the law of conservation, the total angular momentum of the cat must remain zero throughout the fall. But such a state of affairs—falling back first, paws up—clearly does not suit the cat. Then the cat, knowing the law of conservation of angular momentum, finds a simple solution. During the fall, it starts rotating its tail, e.g., clockwise. This causes its body to start rotating counterclockwise. So, to successfully land, the cat rotates its tail until its body turns paws down.

Reader: Amazing! I didn't know I had such a smart cat. Indeed, it rotates, pushing off from its tail. A real cat-physicist. What about the rotation of a helicopter and a sperm cell?

Authors: "It's the same with helicopters. Without a tail rotor, the helicopter's main rotor would spin in one direction, and the helicopter's body in the opposite direction, as required by the law of conservation of angular momentum. The presence of a tail rotor in a helicopter creates a torque acting on the system of the helicopter body—main rotor, and therefore the angular momentum is not conserved, and the helicopter's body does not rotate. But sperm cells lack a tail rotor, and that's precisely why their body rotates in the opposite direction of the flagellum's rotation."

Reader: "Oh, that seems like a clear oversight or mistake by nature. I wonder where Evolution was looking?"

Authors: "No, it's not a mistake. Nature always strives for simplicity and perfection. But let's return to examining the motor section of a sperm cell. The most interesting physics is right here."

"As you might have guessed, the propeller of a sperm cell is powered by a tiny, you could say, nanoelectric motor."

Reader: "Well, what else could it be? It would be absurd if sperm cells trailed a plume of black smoke like a ship with a diesel power plant. Especially since, as I know, for successful fertilization, at least 10 million sperm cells must penetrate into the uterus."

Authors: "Stop, stop! Let's not further develop this thought. You have a rich imagination! Though, if we were to make a comparison, it's better to compare a sperm cell to a submarine. After all, submarines use electric motors."

"As known, a typical electric motor consists of an armature that can rotate, and a stator. Both have windings made of wires through which electric currents flow. In the stator, a magnetic field is created, which acts with Ampère's force on the conductors with current in the armature winding, causing the armature to rotate."

"The rotor of a sperm cell is its middle part, located behind the head, separated by a small constriction called the 'neck.' Behind the middle part is the tail—the propeller. A cytoskeleton of the flagellum, consisting of microtubules, runs through the entire middle part. In the middle part, around the cytoskeleton of the flagellum, is the stator—a mitochondrion, consisting of 28 mitochondria. Remember, mitochondria are the power stations of the cell. The mitochondrion has a spiral shape and wraps around the cytoskeleton of the flagellum. Again, everything looks like in a regular electric motor. The mitochondrion performs the synthesis of ATP and thus provides the electrical energy for the movement of the rotor flagellum. Next, it works somewhat like a myosin motor, but here something (the flagellum) rotates, rather than walking along actin filaments. More precisely, the flagellum performs undulating or helical movements, making 10–40 rotations per second. Energy for the movement of the eukaryotic flagellum is obtained through the hydrolysis of ATP, which is fueled into the sperm cell at its place of origin. With every heartbeat, the male testicles produce approximately 1500 sperm cells."

"It's interesting that thanks to the surface charge located on the membrane, the sperm cell creates an electric field around itself. Observations of mature sperm, performed with an electron microscope, revealed the presence of a dense layer of charges contained in branch-like strands, which create a sufficiently large negative membrane potential."

Reader: "A very delicate question—in which direction does the flagellum rotate, clockwise or counterclockwise?"

Authors: "The flagellum rotates alternately clockwise and counter-clockwise. Scientists haven't fully explored this yet. There might exist chiral sperm cells whose flagellum rotates only in one direction."

Reader: "I have another question. From what you've described, it seems the sperm cell's electric power unit is more advanced than that of an electric car. Can you illustrate this numerically?"

Authors: "It's difficult to compare the range of an electric car and a sperm cell. Their speeds and distances covered are on completely different scales of length and speed. However, in biology, there's a way to compare the speeds of a living organism and a machine, like a leopard and a plane, for example. This involves expressing speed not in meters per second or kilometers per hour, but in body lengths per second. When comparing speeds in such units (number of body lengths per second), one can find out, for instance, that a hummingbird, reaching speeds of up to 27 m/s, flies 385 lengths of its body per second. Meanwhile, a fighter jet only covers 40 lengths of its own per second! So, who flies faster (of course, relative to their sizes and capabilities)? The answer is obvious—the hummingbird. Now, let's do some calculations and compare the range of an electric car and a sperm cell."

"It is known that sperm cells can stay alive and swim for up to 3 days in the cervix, uterus, and fallopian tubes. During this time, moving at an average speed of 30 cm/h, a sperm cell can cover, on average, $3 \times 24 \times 30 = 2160$ cm. Considering that the body length of a sperm cell is 50 μm, we get a range of 21.6 m/50 μm = $4.32 \cdot 10^6$ body lengths. Now, let's do the same calculation for an electric car 4.5 m long with a range of 800 km. We get 800,000/4.5 = $1.8 \cdot 10^5$ body lengths. As you can see, the electro-sperm cell is more than 20 times more advanced than the electric car! But that's not all. We must also consider that the electric car moves in air, while the sperm cell moves in a fluid, where resistance to movement is 800 times greater. When you're at the beach, try running into the water up to your chest. It won't work. There are enough figures and arguments to conclude that the sperm cell's engine is much more advanced and efficient than the engine of a modern electric car."

"By the way, most bacteria use the same principle of movement: electric motor and flagella. Some bacteria have two or more flagella, probably for greater maneuverability when they escape from macrophages eating them. The rotation frequency of bacterial flagella is constant for a specific cell and ranges from 250 to 1700 Hz, i.e., from 15,000 to 100,000 rotations/min! For comparison, in your car, the engine shaft rotates only 1000–5000 rotations/min. As you can see, the bacterial flagellum rotates 10–20 times faster than an automobile engine's shaft."

Diogenes: "From your story, we understand that a sperm cell is an electric machine. But what about the egg cell it so eagerly seeks? Does it also use electricity? How does the egg cell meet the sperm cell?"

Authors: "The egg cell greets the winner of the sperm race with a festive fireworks display, occurring just 200 ns after its membrane makes contact with a sperm cell."

Ralph: "You must be joking?"

Authors: "Not at all. Researchers have been able to witness the mystery of fertilization in action. It turns out that the fertilized egg cell releases a stream of zinc ions (Zn^{2+}), which prevents other sperm cells from penetrating it, strengthening its outer shell. This results in the egg cell glowing—flashes of light—like fireworks, signaling the penetration of a sperm cell. These light flashes are created by zinc ions. The fertilized egg cell puts on this fireworks display and gets rid of excess zinc, which had accumulated during its maturation. Scientists also determined where the egg cell stores the electrical charges (zinc ions) needed for the fireworks and for sealing the membrane against other sperm cells. It accumulates and stores them in special pockets located under the membrane. It has several thousand of these pockets, accumulating about 60 billion zinc atoms. During the fireworks, the pockets fire a volley of about 10 billion 'charges,' making 4–6 volleys."

Ralph: "I understand now. The more intense the fireworks, the better the fertilization result, i.e., the transformation of the egg cell into a full-fledged and healthy embryo."

Diogenes: "Probably so. But I wonder, who is this fireworks display for? Who observes it?"

Authors: "For whom? For the other sperm cells that were late and now swim sadly around."

Reader: "Can I ask a very silly question?"

Authors: "There are no silly questions, only silly answers. Ask away."

Reader: "Do waves form when sperm cells move? And do they move collectively or individually? If waves are formed, other sperm cells might ride the wave created by the one in front. This way, a group of sperm cells could save energy. Do you understand what I'm talking about? For instance, it's known that cranes fly in a V-formation and synchronously (in phase) flap their wings, riding the same wave. In kayak and canoe races, athletes also use the wave-riding effect behind the leader to save energy."

Authors: "Your imagination is running wild today! We understand exactly what you're talking about. Of course, a lone swimming sperm cell creates certain waves in the fluid. These waves were indeed detected by scientists in in vitro experiments and even modeled on a computer. However, we haven't heard about collective phenomena among sperm cell communities. Maybe, after reading this book, you will pursue microbiology and biophysics in the future and discover a new effect in this area. You could name this effect after yourself."

"And to finally convince you of nature's superiority, all we need to add is the fact that sperm cells, unlike electric cars, are very resistant to external influences. For example, mouse sperm cells were exposed to strong cosmic radiation for 6 years while frozen! And after such... It's even hard to find the right word. In short, after this, healthy mouse offspring were born from these irradiated sperm cells: such cute little mice.

Chapter 39

Where in the Universe Can a Cat Warm Up: Entropy, Time, and Life

Dear reader, here you are, having finished this book. We hope that you have not only gained new knowledge about living organisms but also a charge of enthusiasm for seeking answers to new questions abundantly prepared for us by nature. One of the main questions is, what is life? In 1944, the popular science book *What Is Life? The Physical Aspect of the Living Cell* was published, written by the Austrian theoretical physicist Erwin Schrödinger.

Since then, this question has been discussed thousands of times in scientific literature, but scientists have not come to a single definition of life. There are more than a hundred definitions of life. Many of them are quite complex and contradictory. We will present only the simplest and most significant ones from the perspective of physics. For example, according to the definition proposed by NASA in 1994 to find life in the Universe: "life is a self-sustaining chemical system capable of Darwinian evolution." "Life is self-reproduction with variations." A very concise and logical definition.

"Any system capable of replication and mutation is alive." Interestingly, according to this definition (which lacks the word self-reproduction), a virus is a living organism since a virus mutates very

Shadowless Squids: Stories of Physics in Nature
Vitalii Zablotskii and Tatyana Polyakova
Copyright © 2025 Jenny Stanford Publishing Pte. Ltd.
ISBN 978-981-5129-43-4 (Hardcover), 978-1-003-57062-2 (eBook)
www.jennystanford.com

quickly, faster than all other organisms. But, on the other hand, a virus is not a living organism since it cannot exist independently, being just a limited complex of genetic information elements. Viruses are distinguished from living organisms, in particular, because they lack the genetic information necessary for synthesizing the most important systems characteristic of cellular forms of life, e.g., systems responsible for energy production.

Another definition of life points to its source of origin. It is a specific chemical process leading to the emergence of the first self-replicating cells with metabolism that allows them to obtain energy from the surrounding environment. As you can see, a physical quantity—energy—has already appeared here.

One could endlessly discuss all the definitions of life known today, but none of them contains a recipe for creating life. And for a physicist, this sounds strange, as in physics, the definition of any physical quantity contains a recipe for its measurement. For example, at uniform motion, speed equals the path traveled by a point divided by time. If you want to know the speed, measure the path with a ruler and the time with a clock, then divide the former by the latter. And so, it goes with any definition in physics. In the life sciences, there is no quantitative definition of life, i.e., based on measurable physical or chemical quantities. However, there is one, which appeared quite recently.

This definition of life is based on the concept of entropy production—the increase in entropy in a physical system per unit of time as a result of ongoing non-equilibrium processes. The Sun's specific entropy production is about 10^{-4} $W{\cdot}m^{-3}{\cdot}K^{-1}$. The average metabolic rate of specific energy in living matter (from bacteria to anthropoid apes) lies approximately in the range of 0.1–10 W/kg, which for standard conditions and a quasi-stationary process corresponds to a specific entropy production in the range of 0.1–10 $W{\cdot}m^{-3}{\cdot}K^{-1}$. Comparing these quantities, as a hypothesis, the following definition of life can be proposed: Life is a region of space-time with values of specific entropy production in the range of 10^3–10^5 of the specific entropy production of the star near which this region is located. In other words, for living matter, the specific production of entropy is roughly 100,000 times greater than the specific production of entropy of a star.

And one more stroke to the portrait of living matter. Let's compare the specific emission of electromagnetic (thermal) energy by the Sun and a human. This is easily calculated knowing the mass of the Sun ($2 \cdot 10^{30}$ kg) and the energy it emits in one second, $\approx 4 \cdot 10^{26}$ J/s. From this, we get the amount of energy emitted by the Sun per unit of mass per second: $\varepsilon_S = 2 \cdot 10^{-4}$ W/kg. A similar calculation can easily be done for a human, assuming that a person weighing 70 kg consumes (and thus emits, if they are not on a diet for weight loss) 2700 kcal = $1.13 \cdot 10^7$ J of energy per day. Thus, dividing this value by the mass of a person, we get $\varepsilon_H = 2.15$ W/kg. It turns out that a human emits 10,000 times more energy than the Sun (of course, per unit of mass). So who says that stars and the sun shine? Compared to a human, they just smolder. And we are burning, even if we do not emit light in the visible range of electromagnetic waves. And in absolute units, in one year, all humanity produces and consumes $6 \cdot 10^{20}$ J of energy, while only four times more energy from the Sun falls on the surface of our planet: $2.5 \cdot 10^{21}$ J.

Diogenes: "I really like that a human emits 10,000 times more heat than the Sun."

Ralph: "Of course, that's the heat emitted by a unit mass of a human and the Sun."

Diogenes: "Let's not nitpick. Everyone knows we, cats, love to sleep on the head or stomach of our owner, where as much as 2.15 W/kg of heat is emitted, unlike the Sun."

Ralph: "Are you saying that cats know where the warmest places in the Universe are? And it's the human stomach."

Diogenes: "Exactly. We know this clearly without any calculations."

Ralph: "But there's a huge specific entropy production there. And entropy, as is known, is a measure of disorder. Diogenes, does this not scare you?"

Diogenes: "Not at all. I recently caused some disorder myself, knocking a large vase of flowers off the shelf. And I was a bit scared when the owner picked up the broom. I hid under the sofa for a bit, and then everything calmed down."

Authors: "Allow us to continue the conversation about life. We chose this definition of life for a reason. The fact is that life unfolds (evolves) over time, and time and entropy are closely related concepts. Indeed, entropy determines the direction of time (it's like the 'arrow of time') because entropy tends to increase over time. Our Universe is constantly evolving from a state with lower entropy to a state with higher entropy, i.e., a higher degree of disorder. This tendency of a physical system to increase entropy explains why some processes can go forward in time but not the reverse: coffee easily mixes with water, but it's extremely difficult to separate them again. Living organisms or operating machines also constantly increase disorder, e.g., by emitting heat, which increases the entropy of the environment. But unlike machines and natural objects that emit heat, living organisms, per unit mass, produce millions of times more entropy (disorder), thus making the arrow of time so powerful that no one can turn it back! 'The wheel of Time moves indiscriminately,' Krishna spoke."

Reader: "I understand that entropy is a measure of disorder in a physical system. From the perspective of the definition of life based on the production of entropy, an ordinary star produces (per unit mass) hundreds of thousands of times less disorder than any living organism. So what does this mean? Is life just a gigantic source of disorder in the Universe?"

Authors: "From a physics standpoint, you are absolutely right. The Second Law of Thermodynamics states: any isolated physical system tends toward an equilibrium state in which its entropy reaches a maximum. In the 20th century, when developing general theoretical questions of thermodynamics, the principle of maximum entropy production was formulated. The essence of this principle is that a non-equilibrium system evolves in such a way as to maximize its production of entropy given certain external constraints. As you can see, living systems amazingly follow this principle. In biology, a similar principle was proposed by A. Lotka (1922): those species of living organisms that best (all else being equal) utilize portions of the flow of available energy (e.g., solar) for growth and existence, will increase their distribution and numbers."

"So, as we see, all definitions of life are quite general, and therefore they are unlikely to help anyone construct the simplest life from elementary building blocks. However, knowledge of the genetic code of the simplest organisms is simply necessary for this."

Is movement a distinctive feature of life?

We've already mentioned that proteins, built from amino acids, sense the external world, perform work, enable organism movement, and even think. The latter might be an overly bold claim. But that proteins carry out motor functions and facilitate movement is certain.

Note that movement is one of the simplest and most significant indicators of life. Whole organisms, cells, cellular organelles, and intracellular transport all exhibit movement. For instance, the average person walks about 400,000 km in their lifetime. That's more than the distance from Earth to the Moon—380,000 km.

The cause of any movement is a gradient of something essential for life. Indeed, in a homogeneous world, there's no reason for movement, making the phenomenon of life seem impossible. Fortunately, our world is locally heterogeneous, and it's quite remarkable that just a few fundamental gradients determine the life of unicellular and multicellular organisms. Indeed, all known microscopic and macroscopic movements of organisms arise due to a gradient—a change in some quantity's value (e.g., concentration, temperature, light intensity, etc.). The direction of these movements is related to the direction of the respective gradient. For those who have forgotten or don't know what a gradient is: in mathematics, "gradient" is a vector that indicates the direction of the greatest increase of a certain quantity (function) and is equal in magnitude to the rate of increase of this quantity in that direction. For example, if the quantity in question is the height of a mountain above sea level, its gradient at any point on the mountain's surface will indicate "the direction of the steepest ascent" and characterize the steepness of the slope by its magnitude.

Let's consider biological phenomena defined by gradients. It turns out, just six gradients of fundamental physical quantities determine the possible movements of living cells and organisms (Fig. 39.1).

Now let's talk in more detail about these gradients.

(a) The gradient of potential energy of a body's interaction with Earth (gravity) determines the direction of plant growth, early embryo development, and the preferred direction of movement of terrestrial and aquatic organisms.

(b) A temperature gradient determines the direction of thermotaxis (thermotaxis is the behavior in which an organism directs its movement parallel or opposite to the temperature gradient).

Figure 39.1 The development and movement of cells are directed by the following gradients: (a) gradient of the Earth's gravitational potential (gravity), (b) temperature gradient, (c) light intensity gradient, (d) concentration gradient, (e) gradient of the electric potential in cell membrane, and (f) magnetic field gradient.

(c) The gradient of light intensity is responsible for the phenomenon known as phototaxis (phototaxis is considered positive if the movement is toward increasing light intensity, and negative if the direction is opposite).

(d) The gradient of concentration of molecules, ions, and dissolved substances causes chemotaxis—movement in response to chemoattractants, chemorepellents, and diffusion.

(e) The gradient of electric field potential across the cellular membrane establishes the necessary value of the cell's membrane potential.

(f) The gradient of the magnetic field, which is unjustly neglected, probably because the Earth's magnetic field around us is too small and sufficiently uniform at the Earth's surface, meaning the gradient of our planet's magnetic field is practically zero. However, non-uniform (gradient) magnetic fields created by humans surround us in everyday life, as well as in laboratories and technology.

If you look closely again at these six gradients (Fig. 39.1), you'll see that in all cases, the causes of movement are the gradients and the forces they generate. However, these gradients not only determine the directions of organisms' movements but also control nearly all intracellular processes.

Ralph: "Can life and movement exist where there are no gradients, e.g., in a vast homogeneous pink jelly far from stars and planets?"

Authors: "Just imagine for a moment that you're sitting (or rather floating) in a uniform pink yogurt and everything you can see around is also the same pink yogurt. It's just like in the famous song: 'And all that I can see is just another lemon tree. I'm turning my head up and down I'm turning, turning, turning, turning, turning around. And all that I can see is just a yellow lemon tree....' So, everything is absolutely the same, there are no gradients. Or rather, the gradient of any quantity in the surrounding environment is zero. This means the area of space where you're currently floating is no different from a neighboring area or a very distant area. In such a case, do you need to strive and move somewhere?"

Ralph: "I understand. If everything around is absolutely the same, then there's no point in moving anywhere."

Diogenes: Ugh... And I imagined an infinite 3D space, completely filled with identical couches and inexhaustible sources of the most delicious sour cream. And it turned out that there really is nothing to strive for, the couch next door is no better.

Reader: How can you say there's nothing to strive for? For example, from physics, we know that any physical system tends to a state in which its energy is minimal.

Diogenes: You are right. All physical systems strive for a minimum of energy, but people strive for a maximum of money.

Ralph: Fortunately, not everyone. Some people strive for the maximum of knowledge, make remarkable discoveries, and become great scientists.

Authors: Well said, Ralph! We must strive for knowledge and uncover the mysteries of nature. So, friends, forward to the peaks of knowledge, along the gradient of knowledge.

References

References to the selected stories that are recommended for further reading

A Lakeside Concert

1. de Sa FP, Zina J, Haddad CFB. Sophisticated communication in the Brazilian torrent frog *Hylodes japi*. *PLoS ONE*, 11(1), e0145444 (2016). DOI: 10.1371/journal.pone.0145444

2. Buchanan M. Going into resonance. *Nature Physics*, 15, 203 (2019). DOI: 10.1038/s41567-019-0458-z

3. Halfwerk W, Jones PL, Taylor RC, Ryan MJ, Page RA. Risky ripples allow bats and frogs to eavesdrop on a multisensory sexual display. *Science*, 343, 413–416 (2014). DOI: 10.1126/science.1244812

A Play: Conversation About Feelings

1. Kvittingen L, Sjursnes BJ, Schmid R. Limonene in citrus: a string of unchecked literature citings? *Journal of Chemical Education*, 98(11), 3600–3607 (2021).

The Magic of Magnetism

1. Fabricant A, Iwata GZ, Scherzer S, et al. Action potentials induce biomagnetic fields in carnivorous Venus flytrap plants. *Scientific Reports*, 11, 1438 (2021).

2. Lv Y, Fan Y, Tian X, Yu B, Song C, Feng C, Zhang L, Ji X, Zablotskii V, Zhang X. The anti-depressive effects of ultra-high static magnetic field. *Journal of Magnetic Resonance Imaging*, 56(2), 354–365 (2022). DOI: 10.1002/jmri.28035

3. Qin S, Yin H, Yang C, et al. A magnetic protein biocompass. *Nature Materials*, 15, 217–226 (2016). DOI: 10.1038/nmat4484

4. Hand E. Maverick scientist thinks he has discovered a magnetic sixth sense in humans. DOI: 10.1126/science.aaf5803

5. Cooper A, Turney CSM, Palmer J, et al. A global environmental crisis 42,000 years ago. *Science*, 371, 811–818 (2021). DOI: 10.1126/science.abb8677

Mysteries of Bats

1. Hiryu S, Bates ME, Simmons JA, Riquimaroux H. FM echolocating bats shift frequencies to avoid broadcast–echo ambiguity in clutter. *Proceedings of the National Academy of Sciences*, 107, 7048–7053 (2010).

2. Kevin H, Guillerme T, Finlay S, et al. Ecology and mode-of-life explain lifespan variation in birds and mammals. *Proceedings of the Royal Society B*, 281, 20140298 (2014).

Where Hides the Living Electricity

1. Huxley JS, de Beer GR. *The Elements of Experimental Embryology* (Cambridge University Press, 2015).

2. Hall LT, Hill CD, Cole JH, et al. Monitoring ion-channel function in real time through quantum decoherence. *Proceedings of the National Academy of Sciences of the United States of America*, 107, 18777–18782 (2010).

3. Binggeli R, Weinstein RC. Membrane potentials and sodium channels: hypotheses for growth regulation and cancer formation based on changes in sodium channels and gap junctions. *Journal of Theoretical Biology*, 123(4), 377–401 (1986).

4. Zablotskii V, Polyakova T, Dejneka A. Controlling cell membrane potential with static nonuniform magnetic fields. In: Zhang X. (eds) *Biological Effects of Static Magnetic Fields* (Springer, Singapore, 2023). DOI: 10.1007/978-981-19-8869-1_5

A Magical Journey Inside a Living Cell

1. Shiroguchi K, Kinosita K. Myosin V walks by lever action and brownian motion. *Science*, 316, 1208–1212 (2007).

Magnetic Augmented Reality System of a Fox

1. Xie C. Searching for unity in diversity of animal magnetoreception: from biology to quantum mechanics and back. *The Innovation*, 3 (2022). DOI: 10.1016/j.xinn.2022.100229

2. Červený J, Begall S, Koubek P, Nováková P, Burda H. Directional preference may enhance hunting accuracy in foraging foxes. *Biology Letters*, 7, 355–357 (2011).

Champions of Regeneration and Magnets

1. Grohme M, Schloissnig S, Rozanski A, et al. The genome of *Schmidtea mediterranea* and the evolution of core cellular mechanisms. *Nature*, 554, 56–61 (2018). DOI: 10.1038/nature25473

2. Shomrat T, Levin M. An automated training paradigm reveals long-term memory in planarians and its persistence through head regeneration. *Journal of Experimental Biology*, 216, 3799–3810 (2013).

3. Van Huizen AV, Morton JM, Kinsey LJ, et al. Weak magnetic fields alter stem cell-mediated growth. *Science Advances*, 5(1), eaau7201 (2019). DOI: 10.1126/sciadv.aau7201

4. Murugan NJ, Karbowski LM, Lafrenie RM, Persinger MA. Temporally-patterned magnetic fields induce complete fragmentation in planaria. *PLOS ONE*, 8, e61714 (2013).

A Touch of Mysticism: Memory, Hypnosis, and Wrinkles in Time

1. Sadeghi S, Wittmann M, De Rosa E, Anderson AK. Wrinkles in subsecond time perception are synchronized to the heart. *Psychophysiology*, 60, e14270 (2023).

2. Faerman A, Bishop JH, Stimpson KH, et al. Stanford hypnosis integrated with functional connectivity-targeted transcranial stimulation (SHIFT): a preregistered randomized controlled trial. *Nature Mental Health*, 2, 96–103 (2024).

3. Gubin VD. Memory and time. *Vestnik of Saint Petersburg University. Philosophy and Conflict Studies*, 35(2), 358–368 (2019) (In Russian).

Bees that Know Electricity and Magnetism

1. Gould JL, Kirschvink JL, Deffeyes KS. Bees have magnetic remanence. *Science*, 201, 1026–1028 (1978).

Electromagnetic Trash

1. Molina-Montenegro MA, Acuña-Rodríguez IS, Ballesteros GI, et al. Electromagnetic fields disrupt the pollination service by honeybees. *Science Advances*, 9, eadh1455 (2023). DOI: 10.1126/sciadv.adh1455

2. Mian OS, Li Y, Antunes A, Glover PM, Day BL. On the vertigo due to static magnetic fields. *PLoS ONE*, 8, e78748 (2013).

3. Oliveira FTP, Diedrichsen J, Verstynen T, Duque J, Ivry RB. Transcranial magnetic stimulation of posterior parietal cortex affects decisions of hand choice. *Proceedings of the National Academy of Sciences*, 107, 17751-17756 (2010).

4. Cooper A, Turney CSM, Palmer J, et al. A global environmental crisis 42,000 years ago. *Science*, 371, 811-818 (2021). DOI: 10.1126/science.abb8677

Collective Intelligence Living in Magnetic Homes

1. Fleischmann PN, Grob R, Müller VL, Wehner R, Rössler W. The geomagnetic field is a compass cue in cataglyphis ant navigation. *Current Biology*, 28(9), 1440-1444.e2 (2018).

2. Pereira MC, Carvalho Guimarães I, Acosta-Avalos D, Antonialli WFJ. Can altered magnetic field affect the foraging behaviour of ants? *PLOS ONE*, 14(11), e0225507 (2019).

3. Grob R, Müller VL, Grübel K, Rössler W, Fleischmann PN. Importance of magnetic information for neuronal plasticity in desert ants. *Proceedings of the National Academy of Sciences*, 121(8), e2320764121 (2024). DOI: 10.1073/pnas.2320764121

The Price of Immortality

1. Boero F. Everlasting life: the 'immortal' jellyfish. *The Biologist*, 63(3), 16-19.

2. Bavestrello G, Sommer C, Sarà M. Bi-directional conversion in *Turritopsis nutricula* (Hydrozoa). *Aspects of Hydrozoan Biology, Scientia Marina*, 56(2-3), 137-140 (1992).

3. Schmid V, Wydler M, Alder H. Transdifferentiation and regeneration in vitro. *Developmental Biology*, 92(2), 476-488 (1982).

4. Piraino S, Boero F, Aeschbach B, et al. Reversing the life cycle: medusae transforming into polyps and cell transdifferentiation in *Turritopsis nutricula* (Cnidaria, Hydrozoa). *Biological Bulletin*, 190(3), 302-312 (1996).

5. Boero F. Ontogeny, *AccessScience* (McGraw-Hill Education, 2014).

6. Kubota S. Turritopsis sp. (Hydrozoa, Anthomedusae) rejuvenated four times. *Bulletin of the Biogeographical Society of Japan*, 64, 97-99 (2009).

7. Piraino S, Vito DD, Schmich J, et al. Reverse development in Cnidaria. *Canadian Journal of Zoology*, 82(11), 1748–1754 (2004).

How to Jump into a Flying Airplane

1. Książkiewicz Z, Roszkowska M. Experimental evidence for snails dispersing tardigrades based on *Milnesium inceptum* and *Cepaea nemoralis* species. *Scientific Reports*, 12, 4421 (2022). DOI: 10.1038/s41598-022-08265-2

2. Chiba T, Okumura E, Nishigami Y, Nakagaki T, Sugi T, Sato K. Caenorhabditis elegans transfers across a gap under an electric field as dispersal behavior. *Current Biology*, 33(13), 2668–2677.e3 (2023). DOI: 10.1016/j.cub.2023.05.042

3. Shatilovich A, Gade VR, Pippel M, et al. A novel nematode species from the Siberian permafrost shares adaptive mechanisms for cryptobiotic survival with *C. elegans* dauer larva. *PLOS Genetics*, 19(7), e1010798 (2023). DOI: 10.1371/journal.pgen.1010798

Who Flies Without an Engine

1. Williams HJ, Sheparda ELC, Holtonc MD, et al. Physical limits of flight performance in the heaviest soaring bird. *Proceedings of the National Academy of Sciences*, 117(30), 17884–17890 (2020). DOI: 10.1073/pnas.1907360117

Spider on a Flying Carpet

1. Nyffeler M, Birkhofer K. An estimated 400–800 million tons of prey are annually killed by the global spider community. *Scientific Reports*, 104, 30 (2017). DOI: 10.1007/s00114-017-1440-1

2. Morley EL, Gorham PW. Evidence for nanocoulomb charges on spider ballooning silk. *Physical Review E*, 102, 012403 (2020).

3. Han SI, Astley HC, Maksuta DD, Blackledge TA. External power amplification drives prey capture in a spider web. *Proceedings of the National Academy of Sciences*, 116, 12060 (2019).

4. Zhou J, Lai L, Menda G, Miles RN. Outsourced hearing in an orb-weaving spider that uses its web as an auditory sensor. *Proceedings of the National Academy of Sciences*, 119, e2122789119 (2022).

5. Zeng Y, Crews S. Biomechanics of omnidirectional strikes in flat spiders. *Journal of Experimental Biology*, 221 (2018).

6. Redd MA, Scheuer SE, Saez NJ, et al. Therapeutic inhibition of acid sensing ion channel 1a recovers heart function after ischemia-reperfusion injury. *Circulation*, 144(12), 947–950 (2021).

The Marriage of the Chiral Goat

1. Ishimoto K, Ikawa M, Okabe M. The mechanics clarifying counterclockwise rotation in most IVF eggs in mice. *Scientific Reports*, 7, 43456 (2017). DOI: 10.1038/srep43456

2. Yang X, Li Z, Polyakova T, Dejneka A, Zablotskii V, Zhang X. Effect of static magnetic field on DNA synthesis: The interplay between DNA chirality and magnetic field left-right asymmetry. *FASEB BioAdvances*, 2, 254–263 (2020). DOI: 10.1096/fba.2019-00045

Flounder: An Artist and Pharmacist

1. Arnheim R. *Art and Visual Perception* (University of California Press, Berkeley, 1974).

The Largest Organism on Earth

1. Anderson JB, Bruhn JN, Kasimer D, et al. Clonal evolution and genome stability in a 2500-year-old fungal individual. *Proceedings of the Royal Society B: Biological Sciences*, 285, 20182233 (2018). DOI: 10.1098/rspb.2018.2233

2. Nakagaki T, Yamada H, Tóth Á. Maze-solving by an amoeboid organism. *Nature*, 407, 470 (2000).

3. Adamatzky A. On spiking behaviour of oyster fungi *Pleurotus djamor*. *Scientific Reports*, 8, 7873 (2018).

4. Gow NAR, Morris BM. The electric fungus. *Botanical Journal of Scotland*, 47, 263–277 (1995).

5. Adamatzky A. Fungi anaesthesia. *Scientific Reports*, 12, 340 (2022).

6. Adamatzky A. Language of fungi derived from their electrical spiking activity. *Royal Society Open Science*, 9, 211926 (2022).

7. Joshi S, Cook E, Mannoor MS. Bacterial nanobionics via 3D printing. *Nano Letters*, 18, 7448–7456 (2018).

The Oldest Electric Microcheetah

1. Gambelli L, Isupov MN, Conners R, et al. An archaellum filament composed of two alternating subunits. *Nature Communications*, 13, 710 (2022). DOI: 10.1038/s41467-022-28337-1

2. Herzog B, Wirth R. Swimming behavior of selected species of archaea. *Applied and Environmental Microbiology*, 78(6), 1670–1674 (2012).

Frostbitten Crocodile with Flat Eyes

1. Yatsu R, Miyagawa S, Kohno S, et al. TRPV4 associates environmental temperature and sex determination in the American alligator. *Scientific Reports*, 5, 18581 (2016). DOI: 10.1038/srep18581

Extreme Bear: Questions Without Answers

1. Guidetti R, Rizzo AM, Altiero T, Rebecchi L. What can we learn from the toughest animals of the Earth? Water bears (tardigrades) as multicellular model organisms in order to perform scientific preparations for lunar exploration. *Planetary and Space Science*, 74, 97–102 (2012).

Street Quiz: Elephant in Questions and Answers

1. Deiringer N, Schneeweiß U, Kaufmann LV, et al. The functional anatomy of elephant trunk whiskers. *Communications Biology*, 6, 591 (2023). DOI: 10.1038/s42003-023-04945-5

Underwater Electro-Hunting

1. Hofmann BA, Schreyer SB, Biswas S, et al. An arrowhead made of meteoritic iron from the late Bronze Age settlement of Mörigen, Switzerland and its possible source. *Journal of Archaeological Science*, 157, 105827 (2023). DOI: 10.1016/j.jas.2023.105827

2. Gallant JR, Traeger LL, Volkening JD, et al. Genomic basis for the convergent evolution of electric organs. *Science*, 344(6191), 1522 (2014).

3. Sakaki S, Ito R, Abe H, et al. Electric organ discharge from electric eel facilitates DNA transformation into teleost larvae in laboratory conditions. *PeerJ*, 11, e16596 (2023). DOI: 10.7717/peerj.16596

These Mysterious Cats: An Interview with a Renowned Philosopher

1. Jacobs P. How do cats purr? New finding challenges long-held assumptions. *Science*, News, 2023. DOI: 10.1126/science.adl1764

Snakes: Hyper-sensitive and Insensitive

1. Sichert AB, Friedel P, van Hemmen JL. Snake's perspective on heat: reconstruction of input using an imperfect detection system. *Physical Review Letters*, 97, 068105 (2006). DOI: 10.1103/PhysRevLett.97.068105

The Seventh Sense and Earthquakes

1. Logan JM. Animal behaviour and earthquake prediction. *Nature*, 265, 404–405 (1977). DOI: 10.1038/265404a0

2. Boore DM. The motion of the ground in earthquakes. *Scientific American*, 237, 68–78 (1977). DOI: 10.1038/scientificamerican1277-68

3. King C.-Y. Earthquake prediction: Electromagnetic emissions before earthquakes. *Nature*, 301, 377 (1983). DOI: 10.1038/301377a0

4. Grant RA, Halliday T, Balderer WP, et al. Ground water chemistry changes before major earthquakes and possible effects on animals. *International Journal of Environmental Research and Public Health*, 8, 1936–1956 (2011).

5. Homola P, Marchenko V, Napolitano A, et al. Observation of large scale precursor correlations between cosmic rays and earthquakes with a periodicity similar to the solar cycle. *Journal of Atmospheric and Solar-Terrestrial Physics* (2023). DOI: 10.1016/j.jastp.2023.106068

About the Senses of Sharks

1. Keller BA, Putman NF, Grubbs RD, Portnoy DS, Murphy TP. Map-like use of Earth's magnetic field in sharks. *Current Biology*, 31(13), 2881–2886.e3 (2021). DOI: 10.1016/j.cub.2021.03.103

2. Bellono NW, Leitch DB, Julius D. Molecular tuning of electroreception in sharks and skates. *Nature*, 558(7708), 122–126 (2018). DOI: 10.1038/s41586-018-0160-9

3. Knaub JL, Passerotti MS, Natanson LJ, Meredith TL, Porter ME. Vertebral morphology in the tail-whipping common thresher shark, *Alopias vulpinus*. *Royal Society Open Science*, 11(1) (2024). DOI: 10.1098/rsos.231473

Electric Car versus Sperm Cell

1. Calzada L, Salazar EL, N. Pedrón. Presence and chemical composition of glycoproteic layer on human spermatozoa. *Archives of Andrology*, 33(2), 87–92 (1994). DOI: 10.3109/01485019408987808

2. Pacak P, Kluger C, Vogel V. Molecular dynamics of JUNO-IZUMO1 complexation suggests biologically relevant mechanisms in fertilization. *Scientific Reports*, 13(1), 20342 (2023). DOI: 10.1038/s41598-023-46835-0

3. Duncan FE, Que EL, Zhang N, Feinberg EC, O'Halloran TV, Woodruff TK. The zinc spark is an inorganic signature of human egg activation. *Scientific Reports*, 6, 24737 (2016). DOI: 10.1038/srep24737

Where in the Universe Can a Cat Warm Up: Entropy, Time, and Life

1. Martyushev LM. Life defined in terms of entropy production: 20th century physics meets 21st century biology. *BioEssays*, 42(9), 2000101 (2020). DOI: 10.1002/bies.202000101

2. Zablotskii V, Polyakova T, Dejneka A. Cells in the non-uniform magnetic world: how cells respond to high-gradient magnetic fields. *BioEssays*, 40(8), 1800017 (2018). DOI: 10.1002/bies.201800017

Index